Putting Social Movements

Explaining Opposition to
United States, 2000–2005

MW00443406

The field of social movement studies has expanded dramatically throughout the past three decades. But as it has done so, its focus has become increasingly narrow and "movement-centric." When combined with the tendency to select successful struggles for study, the conceptual and methodological conventions of the field conduce to a decidedly Ptolemaic view of social movements: one that exaggerates the frequency and causal significance of movements as a form of politics.

This book reports the results of a comparative study, not of movements, but of twenty communities earmarked for environmentally risky energy projects. In stark contrast to the central thrust of the social movement literature, the authors find that the overall level of emergent opposition to the projects has been very low, and they seek to explain that variation and the impact, if any, it had on the ultimate fate of the proposed projects.

Doug McAdam is Professor of Sociology at Stanford University and the former Director of the Center for Advanced Study in the Behavioral Sciences (CASBS). He is the author or coauthor of thirteen books and some seventy-five articles in the area of political sociology, with a special emphasis on the study of social movements and revolutions. Among his best-known works are *Political Process and the Development of Black Insurgency, 1930–1970*, a new edition of which was published in 1999; *Freedom Summer* (1988), which was awarded the 1990 C. Wright Mills Award and was a finalist for the American Sociological Association's best book prize for 1991; and *Dynamics of Contention* (2001) with Sid Tarrow and Charles Tilly. He is also the author of the forthcoming book, *A Theory of Fields* (with Neil Fligstein). He is a two-time former Fellow of CASBS, a recipient of a Guggenheim Fellowship, and a member of the American Academy of Arts and Sciences (since 2003).

Hilary Schaffer Boudet holds a PhD from the Emmett Interdisciplinary Program in Environment and Resources at Stanford University. Her research interests include the environmental and social impacts associated with energy development and public participation in environmental decision making. Her dissertation focused on the factors and processes that shape community mobilization around proposals for liquefied natural gas facilities. She is currently a postdoctoral scholar at the Stanford University School of Medicine/Stanford Prevention Research Center and a lecturer in the Stanford University Urban Studies Program. She has published in the *Journal of Planning Education and Research*, *Environmental Politics*, *Journal of Construction Engineering and Management*, and *Sociological Forum*.

CAMBRIDGE STUDIES IN CONTENTIOUS POLITICS

Editors

MARK BEISSINGER Princeton University
JACK A. GOLDSTONE George Mason University
MICHAEL HANAGAN Vassar College
DOUG MCADAM Stanford University and Center for Advanced Study
in the Behavioral Sciences
SUZANNE STAGGENBORG University of Pittsburgh
SIDNEY TARROW Cornell University
CHARLES TILLY (d. 2008) Columbia University
ELISABETH J. WOOD Yale University
DEBORAH YASHAR Princeton University

Ronald Aminzade et al., *Silence and Voice in the Study of Contentious Politics*
Javier Auyero, *Routine Politics and Violence in Argentina: The Gray Zone of State
Power*
Clifford Bob, *The Marketing of Rebellion: Insurgents, Media, and International
Activism*
Charles Brockett, *Political Movements and Violence in Central America*
Valerie Bunce and Sharon Wolchik, *Defeating Authoritarian Leaders in
Postcommunist Countries*
Christian Davenport, *Media Bias, Perspective, and State Repression*
Gerald F. Davis, Doug McAdam, W. Richard Scott, and Mayer N. Zald, *Social
Movements and Organization Theory*
Jack A. Goldstone, editor, *States, Parties, and Social Movements*
Tamara Kay, *NAFTA and the Politics of Labor Transnationalism*
Joseph Luders, *The Civil Rights Movement and the Logic of Social Change*
Doug McAdam, Sidney Tarrow, and Charles Tilly, *Dynamics of Contention*
Sharon Nepstad, *War Resistance and the Plowshares Movement*
Kevin J. O'Brien and Lianjiang Li, *Rightful Resistance in Rural China*
Silvia Pedraza, *Political Disaffection in Cuba's Revolution and Exodus*
Eduardo Silva, *Challenging Neoliberalism in Latin America*
Sarah Soule, *Contention and Corporate Social Responsibility*
Yang Su, *Collective Killings in Rural China during the Cultural Revolution*
Sidney Tarrow, *The New Transnational Activism*
Ralph Thaxton, Jr., *Catastrophe and Contention in Rural China: Mao's Great
Leap Forward Famine and the Origins of Righteous Resistance in Da Fo Village*
Charles Tilly, *Contention and Democracy in Europe, 1650–2000*
Charles Tilly, *Contentious Performances*
Charles Tilly, *The Politics of Collective Violence*
Stuart A. Wright, *Patriots, Politics, and the Oklahoma City Bombing*
Deborah Yashar, *Contesting Citizenship in Latin America: The Rise of Indigenous
Movements and the Postliberal Challenge*
Andrew Yeo, *Activists, Alliances, and Anti–U.S. Base Protests*

Putting Social Movements in Their Place

Explaining Opposition to Energy Projects in the United States, 2000–2005

DOUG MCADAM
Stanford University

HILARY SCHAFFER BOUDET
Stanford University

CAMBRIDGE
UNIVERSITY PRESS

CAMBRIDGE UNIVERSITY PRESS
Cambridge, New York, Melbourne, Madrid, Cape Town,
Singapore, São Paulo, Delhi, Mexico City

Cambridge University Press
32 Avenue of the Americas, New York, NY 10013-2473, USA

www.cambridge.org
Information on this title: www.cambridge.org/9781107650312

© Doug McAdam and Hilary Schaffer Boudet 2012

First published 2012

Printed in the United States of America

A catalog record for this publication is available from the British Library.

Library of Congress Cataloging in Publication Data
McAdam, Doug.
Putting social movements in their place : explaining opposition to energy projects in the
United States, 2000–2005 / Doug McAdam, Hilary Boudet.
    p.   cm. – (Cambridge studies in contentious politics)
Includes bibliographical references and index.
ISBN 978-1-107-02066-5 (hardback) – ISBN 978-1-107-65031-2 (paperback)
    1. Energy development – Environmental aspects – United States.   2. Protest movements –
United States.   I. Boudet, Hilary, 1979–   II. Title.
HD9502.U52M3915   2012
333.790973′090511–dc23
                                                                    2011036706

ISBN 978-1-107-02066-5 Hardback
ISBN 978-1-107-65031-2 Paperback

For my earliest mentors, Chick Perrow, Mayer Zald, and especially John McCarthy – Doug

For my husband, Julien Boudet, and my parents, Denise and Daniel Schaffer – Hilary

Contents

Acknowledgments

Over the course of any long-term research project, one invariably collects a host of debts to various individuals who have made contributions to the project along the way. This is our chance to extend our heartfelt thanks to all of these folk. Our biggest debt is to the nearly 250 informants who took time to share their stories and versions of events in their communities with us. In a very real sense, their accounts constitute the empirical foundation upon which this book rests. It is a tribute to our subjects and the time and care they took with their stories that we have as much confidence in this foundation as we do.

We have many academic colleagues to thank as well. The inspiration for the project and the context in which we forged our collaboration was yet another research project. Touched on at various points in the book, the project involved an ambitious interdisciplinary effort to understand the factors that shaped grassroots opposition to water and pipeline projects in the developing world. The linchpin of the project was Ray Levitt, in Stanford's Department of Civil and Environmental Engineering (CEE). But others worked on the project as well, most notably Dick Scott of the Sociology Department and Jenna Davis, also of CEE. We consider ourselves incredibly lucky to have been a part of that collaborative project. Even though we wound up being critical about some methodological aspects of the project, it was the desire to overcome those problems that led directly to our collaboration.

Once it was underway, a host of others aided the project. Pride of place in this regard has to go to the National Science Foundation (SES-0749913), without whose funding support the project would not have been possible. In carrying out the research, we also had the help of a number of talented and dedicated graduate and undergraduate research assistants. No one was more important in this regard than Rachel Gong, who worked full time on the project for some nine months. Several others served shorter stints on the project but were no less committed to the study during their period of involvement. These other research assistants included Sookyung Kim, Griffin Matthew, Erin Olivella-Wright, Beth Siegfried, and Sean Weisberg. Then there were the anonymous reviewers for Cambridge who affirmed the value of the manuscript even as they offered a host of invaluable suggestions for how we might improve the book through revision. They should be able to see the imprint of their many comments and suggestions reflected in the final product.

Finally, we close with two very special thank yous. We are deeply grateful to Sid Tarrow for his close and critical reading of the entire manuscript at an especially formative stage in the book's development. It would be hard to overstate just how much the project benefited from his careful review of that early draft. But no one deserves our thanks more than our friend and colleague Rachel Wright. Rachel joined the project very early on and helped to shape all components of the research strategy. In addition, she was primarily responsible for collecting most of the online data that we used to fashion the "before" profiles of our communities. She also conducted the fieldwork for three of our twenty cases. More than anything, she was a sharp critic, consistent source of inspired research suggestions, and a great pleasure to work with.

Stanford, California
June 2011

From Copernicus to Ptolemy and (Hopefully) Back Again

From its modest beginnings during the 1970s and early 1980s, the study of social movements has developed into one of the largest subfields in American sociology.[1] In recent years, scholars in a host of other disciplines or fields – including political science (Beissinger 2001; Bunce 1999; Dalton, Van Sickle, and Weldon 2009; della Porta 1995; Dosh 2009; Kitschelt 1995; Koopmans 1993; Kriesi et al. 1995; O'Brien and Li 2006; Tarrow 2005; Wood 2003; Yashar 2005), organizational studies (Davis and McAdam 2000; Davis et al. 2005; Ingram, Yue, and Rao 2010; Lounsbury 2005; Rao, Monin, and Durand 2003; Strang and Soule 1998; Vogus and Davis 2005), education (Binder 2002; Davies and Quirke 2005; Hallett 2010; Rojas 2006, 2007; Slaughter 1997; Stevens 2001), environmental studies (Aldrich 2008; Rootes 2003; Rucht 1999; Sherman 2011; Vasi 2011), and law and society (Edelman, Leachman, and McAdam 2010; Gustafsson and Vinthagen 2011; Kay 2005; McCann 1994; Pedriana 2006) – have turned increasingly to social movement theory in an effort to better understand the dynamics of conflict and change within their respective scholarly domains. But even as we acknowledge and celebrate the vibrancy of the field, we worry about what we see as

---

[1] We want to be clear from the outset regarding our usage of the term *social movement studies*. We use the term to describe the interdisciplinary and international community of scholars that gradually emerged during the late 1970s and early 1980s around a shared and explicit identification of social movements as the central object of scholarly interest. With this characterization we are not for a minute suggesting that movements had never been studied before this time. We are, however, arguing that movements had never been the defining empirical focus of a specialized field of study, as they were to become with the birth and subsequent growth of social movement studies.

its increasing narrowness. In his 2009 article in *Annual Review of Sociology*, Andrew Walder voices similar concerns, criticizing what he sees as the field's preoccupation with the dynamics of "mobilization" and general disinterest in a host of broader topics, including the traditional focus among political sociologists on the macrolinks between social structure and various forms of political behavior. There are, however, two other sources of "narrowness" in the field that concern us even more than the one identified by Walder. These are

- The field's preoccupation with movement groups and general neglect of other actors who also shape the broader "episodes of contention" in which movements are typically embedded, and
- The overwhelming tendency of scholars to "select on the dependent variable"; that is to study movements – by which we mean successful instances of mobilization – rather than the much broader populations of "mobilization attempts" or "communities at risk for mobilization" that would seem to mirror the underlying phenomenon of interest more closely.

In combination, these emphases conduce to a starkly Ptolemaic view of social movements. Like Ptolemy, who held that the Earth was at the center of the cosmos, today's movement scholars – at least in the United States – come dangerously close to proffering a view of contention that is broadly analogous to the Ptolemaic system, with movements substituting for the Earth as the center of the political universe. We worry that by locating movement actors at the center of analysis and confining studies to successful instances of mobilization, today's scholars seriously exaggerate the frequency and causal potency of movements while obscuring the role of other actors in political contention.

The research described in this book was motivated by a desire to redress these shortcomings, if you will, to argue for a much more Copernican view of contention, in which emergent grassroots activism – to the extent it develops at all – typically plays only a minor role in episodes of community conflict. You will note that the term *social movement* does not appear in the previous sentence. We weren't so much interested in studying movements per se as the variation in emergent collective action within communities at risk for mobilization. Specifically, we report the results of a comparative case study of the extent to which twenty communities, designated as sites for environmentally risky energy projects, mobilized in opposition to the proposal. Our sample was drawn from all the communities designated for new energy projects for which Final Environmental

Impact Statements (EISs) were required and completed between 2004 and 2007. The list of all such projects – and designated communities – was drawn from the CSA Illumina Digests of EISs for the years noted in the preceding sentence. We will take up the methodological specifics of the study in Chapter 2.

Using a mix of analytic narrative, traditional fieldwork, and fuzzy set/ Qualitative Comparative Analysis (fs/QCA) developed by Charles Ragin (2000), we seek to answer five questions. First, looking at "communities at risk" rather than movements per se, how much emergent collective action do we actually see in our twenty cases? Given the tendency to select actual movements for study, we have very little idea what the underlying baseline of local collective action looks like. In addressing this first question, we hope to shed light on this issue. Second, what "causal conditions" appear to explain variation in the level of mobilization in these communities? Third, net of other factors, what influence, if any, does the level of mobilized opposition have on the outcome of the proposed project? Fourth, we are interested in the dynamics by which localized opposition to a particular kind of energy project – liquefied natural gas (LNG) terminals – scaled up to create broader regional movements against the technology. Why did opposition to LNG terminals grow into broader regional movements in some parts of the country, but not others? And why did that opposition never develop into a truly national movement? Finally, reflecting our desire to broaden out the field and draw insights from scholars in related areas of scholarship, we seek to understand how the research reported here can benefit from, while perhaps contributing to, the rich literature on the policy process that has emerged during the past twenty years or so. We devote considerable attention to this last issue in Chapter 6. Before we take up any of these empirical issues, however, we want to begin by situating this work in a critical analysis of the evolution of the field of social movement studies throughout the past thirty to thirty-five years.

"WHAT A LONG, STRANGE TRIP IT'S BEEN"

Even as we criticize the narrow, movement-centric focus of much contemporary scholarship, we can't help but marvel at the vibrancy of the field and celebrate the fact that social movements are now clearly regarded as an important phenomenon for political analysis. It wasn't always so. When the first author headed off to college as an undergraduate in 1969, he was expecting to be able to take a course or two on social movements. Having committed to work in Washington, D.C., for a coalition of California peace churches as part of broader effort to end the draft, McAdam was hoping to gain a more academic perspective on the history and dynamics of social movements as a form of politics. Logically, from his point of view, he first looked in the course catalog under political science, assuming that was where he would be most likely to find a class on social movements. Nothing. The catalog was filled to overflowing with courses on Congress, international relations, the American electoral system, local public administration, and a host of other institutional features of U.S. politics, but there was nothing whatsoever on social movements or other forms of contentious politics. He remembers briefly surveying the course offerings in the history department, but aside from a course or two on much earlier "contentious moments" in U.S. history – for example, the Civil War and the American Revolution – he found nothing that would give him perspective on more contemporary struggles. Knowing nothing about sociology, he did not even think to look at that section to assay the course offerings in that department. (It would not, however, have mattered if he did. The sociological subfield of social movements was still at least a decade away.)

Mildly disappointed, he filled his schedule with other courses and quickly forgot about his interest in the topic. It wasn't until the fall quarter of his sophomore year that he was reminded of his abortive search when he unexpectedly encountered the topic of social movements in a very surprising course context. Having signed up to take a class in abnormal psychology, McAdam was stunned to see that nearly one-quarter of the course was to be devoted to the topic of – you guessed it – social movements. Having always seen himself as reasonably well adjusted, he was surprised, throughout the quarter, to learn that movement participation was viewed not as a form of rational political behavior but as a reflection of aberrant personality types and irrational forms of "crowd behavior." Who knew?!

In point of fact, McAdam had stumbled upon one of the few corners of the academy that devoted any kind of attention to the study of social movements. But as discordant as the perspective was with his lived experience of activism, the psychological view captured the prevailing view of the phenomenon and fit seamlessly with the broader theories of society and politics that were dominant in American social science at the time.

Structural Functionalism, Pluralist Theory, and Collective Behavior

The social sciences in the United States were characterized by a marked theoretical consensus in the quarter century following World War II. The study of social movements, it should be clear, was a minor academic backwater in sociology and psychology. There were no social movement scholars per se; instead movements were seen as but a minor topic embedded within the broader fields of collective behavior and abnormal psychology in sociology and psychology, respectively. What is interesting, however, is how well the general psychological view of movements reflected the dominant understandings of social and political life characteristic of American social science during the postwar period.

1. Structural Functionalism

The dominant perspective on modern society in this period was structural functionalism. Its leading proponent may have been Talcott Parsons (1951, 1971), but there were few dissenting voices in the consensual choir of postwar American sociology. Whether in Parsons highly elaborated – some might say borderline incomprehensible – version or any of the more accessible variants spelled out in the leading textbooks of the day, structural functionalism offered a view of society as orderly, purposive, harmonious, and, for the most part, free of strain and conflict. Societies – at least functional ones like the United States – were analogous to machines. They were comprised of a complex system of interdependent institutional "parts" – for example, economy, family, education, and politics – that worked together to ensure an overall functional social order. Families socialized children into a broadly consensual normative order; schools reinforced that order and provided the skills and knowledge required to command jobs in the economy, and so on. In this sense, societies, like all machines, tended toward a functioning equilibrium. By implication, serious conflict and change were rare in society, akin to a machine malfunction or breakdown. But again, these were seen as exceedingly infrequent events, bracketing long periods of sustained order and social harmony.

2. Pluralism

Like the functionalist account of society, the dominant model of power and politics during this period – pluralism – also stressed order and consensus over change and conflict as the hallmarks of the American political system. The central tenet of the pluralist model was that political power was widely distributed among a host of competing interests rather than concentrated in the hands of any particular group or segment of society. Thus Dahl (1967: 188–9), perhaps the leading proponent of the theory, tells us that, in the United States, "political power is pluralistic in the sense that there exist many different sets of leaders; each set has somewhat different objectives from the others, and each has access to its own political resources, each is relatively independent of the others. There does not exist a single set of all-powerful leaders who are wholly agreed on their major goals and who have enough power to achieve their major goals."

This wide distribution of power serves to tame the political system. The absence of concentrated power is held to ensure the openness and responsiveness of the system and to inhibit the use of force or violence in dealing with political opponents. With regard to the openness of the system, Dahl (1967: 23) writes that "whenever a group of people believe that they are adversely affected by national policies or are about to be, they .... have extensive opportunities for presenting their case and for negotiations that may produce a more acceptable alternative. In some cases, they may have enough power to delay, to obstruct, and even to veto the attempt to impose policies on them." The implication is clear: groups may vary in the amount of power they wield, but no group exercises sufficient power to bar others from participation in the political system.

Once inside the system, groups find that other organized interests are at least minimally attentive to their political preferences. This responsiveness is again a product of the wide distribution of power held to be characteristic of the pluralist system. Groups simply lack the power to achieve their political goals without the help of other contenders. Instead they must be constantly attuned to the goals and interests of other groups if they are to forge the coalitions that are the key to success in such a system.

Effective political action also requires that groups exercise a degree of tactical restraint in their dealings with other interests. Any attempt to exercise coercive power over other groups is seen as a serious tactical mistake. Lacking sufficient power, contenders are dependent on one another for the realization of their goals. Thus according to proponents of the model, the exercise of force is tantamount to political suicide. Parity in power, then, insures not only the openness and responsiveness of the system but its restrained character as well. "Because one center of power is set against another, power itself will be tamed, civilized, controlled and limited to decent human purposes, while coercion.... will be reduced to a minimum" (Dahl 1967: 24). In place of force and violence, the system will "generate politicians who learn how to deal gently with opponents, who struggle endlessly in building and holding coalitions together ... who seek compromises" (Dahl 1967: 329).

Although internally consistent and very much in keeping with the overarching functionalist perspective on society, the pluralist model made social movements a real puzzle. If the U.S. political system possessed the

normatively appealing characteristics that pluralists ascribed to it – openness, responsiveness, and tactical restraint – how then were we to explain
the puzzling phenomenon of social movements? Why would any group
engaged in rational, self-interested political action eschew the advantages
of such an open, responsive, restrained political system? One possible
answer to the question would be that the group in question has simply
made a tactical mistake. Yet the regularity with which social movements
emerge makes it difficult to believe that, as a historical phenomenon, they
represent little more than a consistent strategic error made by countless
groups. There is, however, another answer, fully consistent with the underlying assumptions of pluralism. Movement participants are simply not
engaged in "rational, self-interested political action." Accordingly, their
rejection of the functional "proper channels" of U.S. politics is not seen as
evidence of tactical miscalculation so much as proof that we are dealing
with an altogether different form of collective behavior. The logic is
straightforward. Social movements represent an entirely different form of
behavior from routine politics. The pluralist model, with its emphasis on
compromise in pursuit of rational self-interest provides a parsimonious
explanation for the latter. Social movements are better left, in Gamson's
wonderful phrasing, to the psychologist or "social psychologists whose
intellectual tools prepare them to better understand the irrational"
(Gamson 1990: 133).

We want to make it clear that this stress on the irrationality of movements was not a component of the pluralist perspective. We are not aware
of any explicit theoretical discussion of social movements by the main
proponents of pluralism. Their silence on the topic owes not to any disdain
for the irrationalism of movements but simply to a shared sense that,
within a pluralist system like the United States, movements were typically
unnecessary and generally ineffective. It was those working in the collective behavior tradition who embraced and articulated a conception of
movements as apolitical and as generally reflecting psychological rather
than instrumental political dynamics.

3. Collective Behavior (and a few brave dissenters)

As we have tried to indicate, there was no defined field of social movement
studies until at least the mid- to late 1970s. There were, however, a handful
of brave souls who defied the functionalist/pluralist consensus during the
postwar period to make the study of social conflict and change the focus of
their work. Most of these scholars were either Europeans – such as Eric
Hobsbawm (1959, 1962), E.P. Thompson (1963, and George Rudé

(1959, 1964) – or American social scientists working in Europe. The latter included Sid Tarrow (1967) and, most importantly, Charles Tilly and a host of colleagues (Rule and Tilly 1972; Shorter and Tilly 1974; Snyder and Tilly 1972; Tilly 1964, 1969; Tilly, Tilly, and Tilly 1975). By contrast, the number of scholars who dared to suggest that conflict and change was also ubiquitous in the United States was exceedingly small. C. Wright Mills (1959) stood virtually alone in this regard during the 1950s but would be joined by others – for example, Gamson (1968a, 1968b) and Domhoff (1970) – as the period wore on.

Although only a few of these scholars were Marxists, virtually all of them were engaged in Marxist-inspired work. As such, they were not interested in social movements per se, but rather in the class basis of social conflict, collective action, and social change. The point is their work did not constitute an embryonic field of social movement studies. There was, however, one small group of American social scientists who did see themselves as studying social movements, though only as one specific instance of a more general social form known as *collective behavior*.

More specifically, the term collective behavior referred to a discrete collection of social forms that were held to be unusual and represent ineffectual, even irrational, responses to the breakdown of social order. These forms included crazes, panics, fads, crowd behavior, and social movements and revolutions. Lumping movements and revolutions together with the other behavioral forms in the list betrays the prevailing apolitical view of the phenomena. At a macrolevel, we were told that movements and revolutions did not so much reflect rational challenges to entrenched political and economic authority as they did dysfunctional responses to the breakdown of social order. In this sense, the perspective bears the stark imprint of the broader structural functionalist theory discussed in the preceding text. Movements – and all forms of collective behavior – were held to arise on those rare occasions when rapid social change (e.g., industrialization, urbanization, and war) occasioned a generalized breakdown in social norms and relationships (Lang and Lang 1961; Smelser 1962; Turner and Killian 1957).

Writing at the time, Gusfield captures the essence of the argument and the close connection between the functionalist view of society and the collective behavior perspective. States Gusfield (1970: 9), "we describe social movements and collective action as responses to social change. To see them in this light emphasizes the disruptive and disturbing quality which new ideas, technologies, procedures, group migration, and intrusions can have for people." The imagery should, by now, be familiar:

society is normally stable, orderly, and self-reproducing. Moreover, people benefit from this order as much as the institutions that comprise society. When the comforting normative order is shattered by the kind of "disturbing" changes to which Gusfield alludes, individuals can be expected to react. Rapid social change is stressful because it undermines the normative routines to which people have grown accustomed. Subjectively, this disruption is experienced as "normative ambiguity," which we are told "excites feelings of anxiety, fantasy, hostility, etc." (Smelser 1962: 11). It is these feelings that motivate all forms of collective behavior including the social movement.

Movements emerge in this view as groping, if ineffective, collective efforts to restore social order and the sense of normative certainty undermined by rapid change. As such, they owe more to psychological rather than political or material motivations. This is not to say that movements are unrelated to politics. Smelser explicitly tells us that movements frequently serve to alert policy makers to significant "strains" in society to which they may need to attend. It is significant, however, that the instrumental political dimension of the movement is reserved for policy makers rather than the movement actors. For the latter, participation is seen as little more than a form of collective coping behavior, motivated by a desire to overcome the stress and uncertainty produced by the breakdown of normative order. The underlying psychological, quasitherapeutic basis of movement participation is implicitly acknowledged by Smelser in his discussion of the "generalized beliefs" that underlie collective behavior. He writes,

collective behavior is guided by various kinds of beliefs.... These beliefs differ, however, from those which guide many other types of behavior. They involve a belief in the existence of extraordinary forces – threats, conspiracies, and so forth. – which are at work in the universe. They also involve an assessment of the extraordinary consequences which will follow if the collective attempt to reconstitute social action is successful. The beliefs on which collective behavior is based (we shall call them *generalized beliefs*) are thus akin to magical beliefs. (Smelser 1962: 8; emphasis in original)

Movement participation is thus motivated by a set of unrealistic beliefs that together function as a reassuring myth of the movement's power to address the stressful state of affairs confronting adherents. Movement participants, we are told, "endow themselves ... with enormous power.... Because of this exaggerated potency, adherents often see unlimited bliss in the future if only the reforms are adopted. For if they are adopted, they argue, the basis for threat, frustration, and discomfort will disappear"

(Smelser 1962: 117). The message is clear: if the generalized beliefs that motivate participation represent a wildly inaccurate assessment of the realities confronting the movement, it is only because they function on a psychological rather than a political level. And so it is for the movement as a whole; they may serve as an early warning to rational political actors that something is amiss in the body social, but burdened by fanciful beliefs, the movement isn't to be taken seriously as a political force in its own right.

Before we close this section, it is important that we place the work of collective behavior theorists in historical context. As reactionary as the collective behavior view of movements would seem to be, it would be wrong to read political conservatism into the perspective. On the contrary, most of the leading proponents of the approach ascribed to political values broadly akin to those of the younger generation of scholars who were to birth the field of social movement studies. Both groups were broadly liberal in their political views; it's just that they were focused on very different types of movements. For the younger scholars, the touchstone struggles, as we will see, were the popular progressive movements of the New Left/New Social Movements with which they strongly identified. The situation was very much the reverse for those who edged toward abnormal psychology in their efforts to understand movements. As Gamson wrote in 1975, "part of the appeal of the collective behavior paradigm is its serviceability as an intellectual weapon to discredit mass movements of which one is critical" (1990 [1975]: 133). For the liberal proponents of collective behavior theory, the modal movements to be explained were such repellant political phenomena as Nazism in Germany, Italian fascism, Soviet-style communism, and McCarthyism in the United States. To again quote Gamson, "who could quarrel with an explanation that depicted the followers of a Hitler or Mussolini as irrational victims of a sick society?" (1990 [1975]: 133).

In short, as in all fields of knowledge, political values and contemporary social concerns shaped the scholarship of the collective behaviors theorists and the newer generation of scholars who rejected the perspective in favor of a more explicitly rational political view of movements. More importantly, these concerns and values introduced opposite biases into the study of social movements. If political antipathy to the movements they studied prompted the proponents of collective behavior to stress the irrationality and general ineffectiveness of movements, the strong political identification of the newer generation with their touchstone movements (e.g., civil rights, women's liberation, and peace), encouraged opposite tendencies. The evolution of the field, we will argue in what follows, betrays the biases inherent in this strong, positive identification.

Summing Up

In combination, this mix of theoretical perspectives, painted an entirely consistent portrait of social-political life. Modern society (read: American society) was akin to a well-ordered, functional machine, comprised of an interdependent and mutually reinforcing set of institutional parts. The political system was but one of those parts, but in the felicitous view of the pluralists, it too served to promote social order, stability, and a broad (and appealing) normative consensus. Collective interests had only to organize in order to gain access to and take advantage of the system, which was, by turn, open, responsive, and hard on those who would dare to use force or violence to impose their will on others.

At first blush the very existence of social movements would appear to pose a stark challenge to the central thrust of structural functionalism and pluralist theory. However, collective behavior theory resolves the apparent contradiction. Movements are not anomalous at all; instead, along with the other forms of collective behavior, they represent the rare exception to the rule of order, harmony, and equilibrium that is characteristic of modern society. Movements arise on those rare occasions when the normative order has been undermined by rapid social change, prompting those subject to the change processes to engage in a form of reactive collective behavior. Movements are thus akin to the proverbial canaries in the coal mine, signaling to rational policy makers that some form of political intervention may be necessary in order to restore the normal steady state of the functioning social-political system. The movements, however, are but a psychological expression of the underlying strain in society rather than a meaningful political response to it.

THE SIXTIES: FROM COLLECTIVE BEHAVIOR TO SOCIAL MOVEMENT STUDIES

The tidy, parsimonious view of social-political life fashioned by the structural functionalists and pluralists was no match for the turbulence of the 1960s and 1970s. In truth, the immediate postwar period was never quite as orderly or conformist as the stereotype of the 1950s would have us believe. (Any decade that could give rise to the modern civil rights movement and the shock of rock and roll – think Little Richard and Jerry Lee Lewis at their wildest – is hard to label as conformist!) But compared to what lay ahead, the immediate postwar period did appear to accord with the central thrust of structural functionalism and pluralist theory. The baby boom and supporting pronatalist ethos encouraged a lockstep, life-course conformity never seen before or since in the United States. Everybody did seem to be on the same page. Or at least the dominant cultural images – lovingly reproduced on the pages of *Life* and the *Saturday Evening Post* and propagated even more forcefully through the new medium of television – suggested as much. Politics – notwithstanding the "little" matter of race – did bear a superficial resemblance to the pluralist vision, especially during the Eisenhower years. The policy prefer-ences of Democrats and Republicans were nearly indistinguishable on the three major issues of the day: civil rights, the Soviet threat, and the gradual extension of the welfare state. No major movements (save for the occa-sional unpleasantness regarding race, e.g., Montgomery and Little Rock) disturbed the surface civic calm of American life. If one squinted enough and the lighting was just right, the social scientific stress on order, stability, and consensus appeared to fit the reality of life (at least its middle-class variant) in the postwar United States. Then along came the sixties.

Very quickly the generalized unrest of the 1960s shattered the theoret-ical consensus described in the preceding text. With respect to structural functionalism, the ubiquitous nature of conflict and change during the period made the earlier stress on harmony and stasis seem outdated and naïve. Conflict perspectives began to proliferate in American social science. The turbulence of the 1960s simultaneously undermined scholarly faith in the pluralist account of U.S. politics. As civil rights forces battled intran-sigent segregationists and the Johnson and Nixon administrations appeared to turn a blind eye to a broad-based antiwar movement, the dominant image of American politics shifted from that of an open and responsive political arena to a closed and coercive "power structure." Elite theories of power and politics quickly supplanted a discredited pluralist theory, which, in turn, made the apolitical, psychological view of social movements increasingly untenable as well. If American politics was less an "open arena" than an entrenched "power structure," recourse to psychol-ogy was hardly required in order to understand the impulse to protest. Movements were simply politics by other means; they were rational efforts to generate leverage by groups (e.g., African Americans and antiwar activists) that were denied access to the "proper channels" near and dear to the hearts of pluralists. Working from these assumptions, a new gen-eration of scholars – with sympathies for if not roots in, the popular struggles of the period – began to fashion very different theoretical accounts of the movement phenomenon. In doing so, they had one clear advantage over today's would-be movement scholars: they were not constrained by the received wisdom and theoretical conventions of today's large, well-established, and often-insular field of social movement studies.

Lacking any such narrowly appropriate audience, the pioneering scholars whose work would come to constitute the foundations of the field were encouraged to frame their research in much broader terms and to engage with scholars in a wide range of other fields. Three very different but quite broad framings can be discerned in the earliest "rationalist" writings about movements from this period. We refer to these three traditions as *political process*, *resource mobilization*, and *political economy*.

Political Process

In the narrow U.S.-based sociological account of the development of social movement studies it has long been the convention to assign pride of place to resource mobilization as the first explicit theoretical challenge to the dominant collective behavior perspective. But that is not strictly true and probably reflects a disciplinary bias in the construction of the field's origins. Although not synonymous with American sociology, the field of social movement studies has been substantially rooted there for most of its history. That is probably why resource mobilization and its two sociology progenitors, John McCarthy and Mayer Zald, are typically credited with triggering the paradigm shift that gave birth to the contemporary sociological study of social movements.

In truth, however, two political scientists, Michael Lipsky and Peter Eisinger, were first off the mark in this regard. Both were scholars of urban politics trying to make sense of struggles for power and influence during one of the most tumultuous and violent periods in the history of U.S. cities. In this context, it was clear to both scholars that movements had emerged as a highly visible and consequential form of politics in the United States, which is to say that the traditional collective behavior paradigm was of little help to them in making sense of the role that movements were clearly playing in the urban politics of the era. Instead they sought to understand movements in a dynamic relationship with the systems of institutionalized political power that the movements sought to challenge. In this sense their audience was other political scientists and political sociologists interested in the nature of power and the ways in which state authority shaped the variable fate of popular movements. Although functionalists and pluralists emphasized order and stability in social and political life, Lipsky and Eisinger saw conflict and change and sought to understand the oscillation between the two. Writing in 1970, Lipsky urged political analysts to avoid system characterizations presumably true for all times and places, which are basically of little value in understanding the social and political process. We are accustomed to describing communist political systems as "experiencing a thaw" or "going through a process of retrenchment." Should it not at least be an open question as to whether the American political system experiences such stages and fluctuations? Similarly, is it not sensible to assume that the system will be more or less open to specific groups at different times and at different places? (40)

It is clear from the research reported in his groundbreaking 1970 book *Protest in City Politics* that Lipsky thought the answer to this last question was yes. He clearly believed that the opportunities for a challenger to engage in successful collective action varied over time and that the 1960s constituted one of those periods in which the system was generally more open and vulnerable to challenge by movement groups. That protest, to use his term, had become a significant "political resource" of the traditionally powerless.

In his research on the variable effectiveness of "protest behavior" across a sample of American cities, Eisinger drew on Lipsky's work and extended it. Although Lipsky had stressed temporal variation in receptivity and vulnerability to movement activity, Eisinger focused on geographic variation. More specifically, he sought to analyze the link between the level of protest behavior in his cities and the relative openness of their "political opportunity structures." His results were revealing: protest rates were significantly higher in cities whose elected political officials were sympathetic to the demands of protestors. Eisinger's conclusion anticipated one of the central insights of what would come to be known as the *political process model*. Wrote Eisinger, "Protest is a sign that the opportunity structure is flexible and vulnerable to the political assaults of excluded groups." Put simply, movements tend to develop in times and places where political elites are newly vulnerable or receptive to challenge.

But if the work of Lipsky and especially Eisinger "anticipated" the political process model, it remained for a trio of sociologists, William Gamson, Charles Tilly, and Doug McAdam, to extend and elaborate their insights and formally explicate the perspective. Gamson's work came first, in the form of his groundbreaking 1975 book *The Strategy of Social Protest*. Although Gamson's book was not a theoretical work but was the report of an innovative research project, it nonetheless incorporated any number of insights and concepts that would become central to the political process framework. Like Lipsky and Eisinger before him and Tilly and McAdam to follow, Gamson's focus was squarely on the dynamic, interactive relationship between "challenging groups" and constituted state authority. In the distinction he drew between "members" and "challengers" Gamson anticipated the "polity model" that Tilly formally articulated in *From Mobilization to Revolution* in 1978. As Gamson (1990: 140) explained the distinction: "the central difference among political actors is captured by the idea of being inside or outside of the polity. Those who are inside are *members* whose interest is vested – that is, recognized as valid by other members. Those who are outside are *challengers*. They lack the basic prerogative of members – routine access to decisions that affect them." Finally, in devoting a full chapter to the empirical relationship between "crisis" and movement success, Gamson showed that he too was attuned to temporal variation in the "opportunities" available to challengers to press their claims.

In *From Mobilization to Revolution*, Tilly was far more consciously theoretical and programmatic than Gamson had been in *The Strategy of Social Protest*. Most of the conceptual elements that we now associate with the political process perspective were on display in Tilly's book. Besides the formalization of the polity model implied in Gamson's work, these "conceptual elements" included

- A stress on "opportunity" and "threat" as distinct catalysts of mobilization,
- The critical importance of state facilitation and repression in shaping the dynamics of contention, and
- An explicit treatment of the concept of repertoires.

But it remained for McAdam, writing in 1982, to formally explicate a political process model of social movements. Although drawing on all of the works discussed in the preceding text, McAdam incorporated two additional elements into his explanatory framework. McAdam argued that the "structural potential" for a movement was defined by favorable political opportunities and access to mobilizing structures (established groups or networks) independent of elite control. But "while important, expanding political opportunities and ... organization do not, in any simple sense, produce a social movement.... Together they only offer insurgents a certain objective 'structural potential' for collective political action. Mediating between opportunities and action are people and the subjective meanings they attach to their situation" (McAdam 1999 [1982]: 48). McAdam dubbed the transformation of consciousness that he saw as mediating between opportunity and action "cognitive liberation."

Over time, many more scholars would add to this distinctly political, state-centered, account of movements; perhaps none more so than Sidney

Tarrow (1983, 1989, 1998). But the basic perspective was in place by the early 1980s. Despite considerable criticism of the model (Armstrong and Bernstein 2008; Goodwin and Jasper 1999), it remains arguably the most influential account of movement dynamics in the field.

Resource Mobilization

As noted earlier, McCarthy and Zald were arguably the first sociologists to challenge traditional collective behavior theory explicitly. In doing so, they shared much in common with those who were simultaneously fashioning the political process perspective. Most importantly, all of these scholars rejected the dominant psychological account in favor of a much more straightforward rational, instrumental view of social movements. Gamson (1990: 138–9) spoke for all of the newer generation of scholars when he wrote that "in place of the old duality of extremist politics and pluralist politics, there is simply politics.... Rebellion, in this view, is simply politics by other means. It is not some kind of irrational expression but is as instrumental in its nature as a lobbyist trying to get special favors for his group or a major political party conducting a presidential campaign."

Although Lipsky, Tilly, and the others reviewed in the previous section were intent on embedding research on social movements in the broader study of politics, McCarthy and Zald saw themselves as organizational scholars trying to make sense of the rapid proliferation after World War II of a relatively new kind of organization. The title of their groundbreaking 1973 piece, *The Trend of Social Movements in America: Professionalization and Resource Mobilization*, captures the broad, historical perspective they brought to their initial statement of the resource mobilization perspective. Much of the piece was not about social movements per se but rather was about the broad changes in American society – for example, generalized prosperity and the substantial rise in discretionary income – that had fueled the growth of this particular organizational sector. Although perhaps not quite as broad a frame of reference as the proponents of political process had brought to their work, McCarthy and Zald were not speaking narrowly to social movement scholars. Instead their work was really pitched to the interdisciplinary field of organizational studies.

Attracted by their explicit critique of collective behavior and the starkly realist – some might say faintly economistic – view of movements as just another form of strategic organizational behavior, McCarthy and Zald emerged as the leading figures in the embryonic social movement subfield beginning to take shape in American sociology. As such, resource mobilization quickly attracted a number of adherents, who sought to frame their work in relation to the new perspective (Ferree and Miller 1985; Jenkins and Perrow 1977; Oliver 1984). But as a distinct theoretical perspective, resource mobilization developed very differently than political process. Although the latter developed slowly through an accretion of insights by a number of scholars, with the term *political process* only being formally invoked late in this process, resource mobilization was born fully formed in the earliest writings by McCarthy and Zald. They used the term in their very first piece and followed that up with a formal programmatic statement of the perspective in their classic 1977 article.

We close by noting one other difference between the resource mobilization and political process approaches. Although political process directs attention away from movements to the broader political environments in which these movements are embedded, the focus of resource mobilization is squarely on the movement and especially the formal social movement organizations (SMOs) that are seen as the public face and organizational carriers of the struggle. Interestingly, even as political process has emerged as perhaps the most influential theoretical perspective on social movements, the narrower, movement-centric focus of resource mobilization has come to define the lion's share of work in the field. We will have more to say about this later in the chapter.

Political Economy

For all the attention that resource mobilization and political process have received throughout the years and the influence they continue to exert today, few seem to remember that there was a third perspective on social movements as the field was coming together during the 1970s. If political process emphasized the broader political context and resource mobilization the organizational dynamics of movements, another group of contemporary scholars sought to understand the relationship between movements and revolutions and the systems of economic production and class relations that undergird all of social and political life. In framing their work in this way, proponents of this third approach were clearly drawing on the earlier Marxist-inspired scholarship touched on in the preceding text. The difference is that, unlike their predecessors, the figures working in this third vein saw themselves as scholars of social movements and revolution.

In a recent article, Goodwin and Hetland (2010) have called attention to this largely forgotten third perspective and what they call its "strange disappearance" from social movement studies. To highlight the central explanatory focus on economic factors or processes, the authors briefly survey a number of works in this tradition. There is, for instance, Jeffrey Paige's widely cited 1975 book *Agrarian Revolution*, which sets forth a general theory of how the structure of agricultural production and class relations conditions the likelihood of revolution.[2] A year after Paige's book appeared, Michael Schwartz published his explicitly Marxist account of the rise of the Southern Farmers' Alliance during the 1880s. Schwartz sought to show how the system of cotton tenancy in the South during this period shaped the grievances, organizational structure, and tactical choices of the Southern Farmers' Alliance.

Another work that focused on the role of the southern cotton economy in shaping the political fortunes of a different movement was McAdam's 1982 account of the origins of the modern civil rights struggle. The central importance assigned to economic factors in McAdam's account is captured by the following quote: "if one had to identify the factor most responsible for undermining the political conditions that, at the turn of the [twentieth] century, had relegated blacks to a position of political impotence, it would have to be the gradual collapse of cotton as the backbone of the . . . economy" (1999 [1982]: 73). The decline of "King Cotton" aided the movement in three ways. First, it undermined the material logic of Jim Crow, rendering the system more vulnerable to challenge. Second, the accompanying decline in the need for agricultural labor helped to set in motion the "Great Migration" of African Americans out of the South. The political significance of the exodus owed to the fact that the overwhelming majority of the black migrants settled in the key northern industrial states, greatly enhancing the value of the so-called black vote and the national political significance of civil rights forces. Finally, the decline in the cotton economy also triggered a significant rural-to-urban migration within the South, in the process greatly strengthening the three urban institutions – black churches, black colleges, and National Association for the Advancement of Colored People chapters – that would serve as the locus of movement organizing during the 1950s and 1960s.

---

[2] Goodwin could just as easily have mentioned Paige's much later book, *Coffee and Power* (1997), as another work in this tradition. As in his earlier work, Paige stresses the central role of economic forces – in this case the reaction of each country's "coffee elite" to the political and economic crisis of the 1980s – to explain the very different developmental paths taken by Nicaragua, El Salvador, and Costa Rica during the 1980s and 1990s. But perhaps Goodwin omitted the reference because it didn't accord with his polemical claim that works in the "political economy" tradition have all but disappeared from the study of social movements and contentious politics.

Tilly's work from this same period also reflected a fundamental interest in matters of political economy. The emphasis in Tilly's scholarship, however, was somewhat different than in the previously cited works. In his groundbreaking work on popular contention in nineteenth-century Europe, he made much of the way that industrialization shaped not only the grievances of an emerging proletariat but also more importantly – at least for Tilly – the structures available to the working class within which to mobilize. Although the collective behavior theorists saw industrialization as one of those destabilizing change processes that encouraged protest by producing widespread stress among the affected, Tilly (and his colleagues) emphasized very different explanatory mechanisms. By concentrating large numbers of workers in factories and ecologically dense urban neighborhoods, industrial capitalists unwittingly afforded the proletariat ideal settings within which to mobilize (Tilly, Tilly, and Tilly 1975).[3]

Writing in the same general period, other scholars invoked economic arguments in their accounts of additional movements (Anderson-Sherman and McAdam 1982; Piven and Cloward 1977; Skocpol 1979). One could argue that the stress on political economy was more marked in the study of social movements during this era than was the emphasis on either institutional politics or organizational processes. Yet the former has, as Goodwin and Hetland contend, largely disappeared from social movement theory, while the other two animating frameworks have thrived. Our interest here, however, is not in speculating why resource mobilization and political process survived and political economy did not, or in arguing, as Goodwin does, for its relevance in the case of most movements. Instead we close this section by highlighting what is common to all three of the fledgling perspectives that vied for influence in the early days of the field. Quite simply, all of them sought to embed the study of social movements in a much broader field of study, whether political sociology or science, organizational studies, or research on political economy. These were, to return to the analogy introduced earlier, decidedly Copernican views of the phenomenon in question; they were attempts to understand movements in relationship to much broader social, political, and economic forces.

---

[3] But if Tilly and McAdam incorporated political economy into some of their early groundbreaking works, this focus is decidedly absent from virtually all of their published writings after the 1970s for Tilly and after the mid-1980s for McAdam.

HONEY, I SHRUNK THE FIELD

The study of social movements has been one of the real "growth industries" in the social sciences throughout the past twenty-five to thirty years. This growth has been especially impressive in American sociology in which the field first took root. At present, the American Sociological Association (ASA) recognizes forty-eight sections representing different subfields within the discipline. The rapid expansion of the social movement subfield is nicely captured in Figure 1.1, which shows the number of members in the section on collective behavior and social movements (CBSM) from its inception through 2010.

From twelve members in 1979, the section had grown, through 2010, to be the seventh largest in the ASA. As noted at the outset of the chapter, the field is now expanding rapidly outside of sociology, with considerable resonance in organizational studies, education, political science, and legal studies.

Yet even as scholarship on social movements has proliferated, the substantive focus of that work has grown increasingly narrow. This isn't as strange as it might appear at first blush. As we have argued, the absence of a field of social movement studies forced those scholars whose works defined the emerging area to read widely and frame their work for much broader audiences (e.g., organizational studies and political economy) than contemporary scholars typically do. They simply did not have today's scholar's limiting "luxury" of being able to situate their work in a narrow

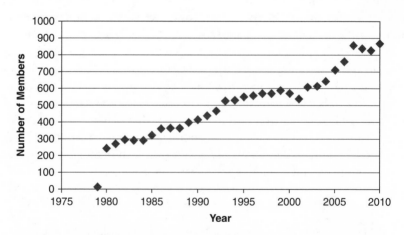

FIGURE 1.1. Membership in the American Sociological Association section on "collective behavior and social movements")

body of social movement theory and research. As the field developed, however, it grew sufficiently large as to constitute its own primary audience, encouraging it to become increasingly insular and self-referential in the process. The establishment of the "specialty" journals *Mobilization* and *Social Movement Studies* has only reinforced this tendency to speak primarily to other social movement scholars.

An examination of the index of perhaps the most prestigious reference volume in the field, *The Blackwell Companion to Social Movements*, affords a telling, if indirect, reflection of the narrowness that has come to characterize social movement studies. Consider the following list of index entries that would seem to speak to a broader contextual understanding of social movements:

Elections/electoral systems – six pages
Economic instability – two pages
Capitalism/capital – five pages
World economy – two pages
World system theory – eight pages
State(s)/state breakdown – forty-nine pages
Political parties – four pages

With the exception of "state(s)/state breakdown," the listings for these various topics are meager. If, at the outset, the field was substantially preoccupied with understanding movements in macropolitical and economic context, this broader "external" focus has all but disappeared from contemporary scholarship. Contrast these paltry numbers with the large number of page listings for the following set of narrowly "internal" topics:

Framing/frames – ninety-six pages
Emotions – thirty pages
Tactics/tactical repertoires – thirty-nine pages
Social movement organizations – forty-eight pages
Collective identity – forty-seven pages
Mobilization – seventy-five pages

The attention devoted to the preceding topics is entirely consistent with the argument advanced by Walder (2009a) in his recent survey of the field. The focus of social movement scholarship is now squarely on mobilization, those who mobilize, and in general, internal movement dynamics.

In short, although the growth of the field may, in general, be cause for celebration, it was achieved, in our view, at considerable intellectual cost. Most importantly, as the area of study coalesced, broader visions of the emerging field gradually gave way to a narrow preoccupation with social movements qua movements. Given the breadth of the work reviewed in the preceding text, this particularly narrow vision of the central object of study was hardly inevitable. Most of the early contributors never identified with the study of social movements per se and so eschewed affiliation with the field and continued to do much broader work for other audiences. This characterization would certainly apply to scholars such as Francis Fox Piven, Richard Cloward, Michael Schwartz, Jeffrey Paige, Theda Skocpol, and even Charles Tilly.

The narrow focus on movements, however, did much more than simply deter some scholars from identifying with and continuing to contribute to the emerging field. There was a much more serious intellectual downside to the creeping narrowness. In focusing primarily on movements, the emerging community of movement scholars began unwittingly to push to the margins the very actors – economic elites, state officials and political parties – that had been central to much of the pioneering work that shaped the field in the first place. Gradually, Ptolemy replaced Copernicus as the guiding spirit of the emerging field. Instead of situating movements in a fuller constellation of political and economic forces and actors, movements and movement groups increasingly came to be the central animating focus of the field. The danger in this is the same embodied in Ptolemy's geocentric model of the cosmos. By locating the Earth at the center of his model, Ptolemy greatly exaggerated its overall importance and seriously distorted the causal dynamics of our solar system. Instead of the sun being the central dynamic force in the system and the properties of our planet as largely shaped by that force, Ptolemy reversed the relationship in his model. By focusing centrally on movements and ignoring or marginalizing other political or economic actors, movement scholars began to unwittingly fashion a conception of conflict processes that was akin to Ptolemy's model of the cosmos. Movements emerged as powerful vehicles of social and political change. Movements were all about agency and causal impact. Movements weren't so much acted upon as acting.

At least three other factors reinforced the tendency to see movements as powerful causal forces in the world. The first was simply the values that motivated most movement analysts. It is not much of an exaggeration to say that the field was born of a sustained interrogation of the movements of the New Left in the United States and the new social movements in Europe. The great majority of the scholars who were active in the early days of the field had either participated in or strongly identified with these movements. It seems reasonable to imagine that this close connection reinforced a general myopia and the tendency to emphasize the social and political importance of movements.

The "cultural turn in the social sciences," which more or less coincided with the development of the field, probably encouraged the Ptolemaic take on movements as well. Although an emphasis on macrostructural forces had been all the rage with Marxists and other conflict theorists during the late 1960s and 1970s, culturalists criticized the "structural determinism" inherent in these approaches and sought instead to emphasize the central importance of human agency and the cultural processes through which this agency was exercised. In the field of social movement studies, this meant a growing antagonism toward broad structural accounts emphasizing the influence of such disembodied forces as political opportunity structures and systems of class oppression, in favor of an emphasis on cultural processes such as framing and emotion work that spoke to the cultural creativity and agency of movement forces. But this often meant organizing other powerful political and economic actors out of the analysis and granting center stage increasingly to movement leaders, activists, and organizers.[4]

But perhaps nothing has served to reinforce the exaggerated importance of movements more than the methodological convention of selecting successful instances of mobilization for study. That is, our theories of movement emergence are almost entirely based on the study of successful instances of mobilization. This carries with it all the standard risks associated with "selecting on the dependent variable." It is very likely that visible, sustained movements are so wildly atypical of mobilization attempts that our understanding of emergence is critically compromised by focusing only on them. The empirical phenomenon of interest should really be something more general such as "mobilization attempts" or better yet "populations at risk for mobilization" rather than the exceedingly rare and almost certainly atypical instances of sustained mobilization that have been the empirical staple of social movement studies. Yet despite some awareness of the issue, movement scholars continue to perpetuate the problem by singling out mature movements for study.

To put the matter as starkly as possible, the conceptual and empirical conventions governing the field are geared to yield "evidence" that affirms

---

[4] Although most cultural analysis of movements have tended to be "movement-centric" in their focus (i.e., focused on cultural processes internal to this or that struggle), there are clear exceptions to this generalization. The work of such culturally attuned scholars as William Sewell Jr. (1992, 1996), Marc Steinberg (1994, 1999), Elisabeth Clemens (1997), and Ann Swidler (1986), among others, consistently seeks to understand cultural dynamics within movements against the backdrop of much broader economic, political, and social processes.

the causal importance and frequency of social movements. By ignoring the role that other powerful actors and forces play in episodes of contention, selecting instances of successful mobilization for study, and attending closely to the "agentic" activities of activists and organizers, movement scholars are virtually assured of "finding" that movements matter. But given the myopia clearly reflected in these conventions, we should be wary of the affirming evidence. Attend too closely to any empirical phenomenon, and you are bound to reify and exaggerate its importance.

WIDENING THE ANGLE OF VISION

The research reported in this book was born of our concern that the myopia reflected in much movement scholarship threatened to distort our understanding of the true frequency and social-political importance of emergent contentious action. We sought to overcome this distorting narrowness through our research design. The basic study is easy to describe. Ours is a comparative case study of twenty communities subject to the "threat" of large, environmentally sensitive energy projects. In its basic design, the study is quite different from and much broader than most social movement scholarship. Three features of our study bear mention.

First, we resolved to study communities at risk for mobilization rather than social movements per se. So instead of researching instances in which local movements developed in opposition to proposed energy infrastructure projects, we picked twenty cases at random to study from the broader population of all communities potentially subject to such projects. The population of such cases was drawn from the register of all energy projects that completed a Final EIS between 2004 and 2007. In substituting these at-risk communities for movements, we sought to avoid the all too common tendency of social movement scholars to select on the dependent variable.

Second, rather than focusing exclusively or primarily on social movement actors, our procedures were geared to identify all of the relevant actors – individuals as well as collectivities – who emerged as important "players" in the decision-making process that determined the fate of the project. By focusing equal attention on all parties to the conflict, we hoped to be able to assess the degree to which any emergent opposition shaped the outcome of the project relative to other significant actors.

Finally, it is worth highlighting the spatial locus of our research. Without denying the many good studies of local movements that have been carried out, the field has nonetheless focused most of its attention on broader national or even transnational social movements. It should be obvious that the disproportionate emphasis on these broader struggles only exacerbates the problem of nonrepresentativeness touched on in the preceding text. If sustained mobilization is a rare occurrence, how much rarer are those cases in which local mobilization scales up to either national or transnational levels? By studying at-risk local populations we hope, once again, to be analyzing more representative instances of mobilization or mobilization attempts.[5]

The design of the study reflects the motivating force of the four specific research questions noted at the outset of the chapter:

1. How much oppositional mobilization do we see across our twenty communities?
2. What "causal conditions" explain variation in the level of mobilization in these communities?
3. Net of other factors, what influence, if any, does the level of mobilized opposition have on the outcome of the proposed projects?
4. Why did opposition to one kind of energy project – LNG terminals – grow into broader regional movements in some parts of the country but not others? And what mechanisms appear to shape this upward "scale shift?"

PLAN FOR THE BOOK

We close the chapter with a brief road map of what is to follow in the rest of the book. In Chapter 2, we detail the mix of empirical methods employed in the comparative study of our communities. At the risk of being immodest, we think the study is almost as important for its innovative research design as for its animating theoretical reorientation of the field. At least three aspects of the project set it apart from the methodological conventions of most social movement research. First, as noted in the preceding text, we choose to study communities at risk for mobilization rather than movements per se. Second, wanting to combine the empirical richness of the traditional case-study approach with the inferential possibilities of large-N statistical studies, we choose to structure the research

---

[5] In their broad empirical overview of "protest events" in Chicago between 1970 and 2000, McAdam et al. (2005: 11) found that only slightly more than 5 percent of all protests were linked to broader national struggles. In contrast, nearly three-quarters of all such events were focused on either city- or neighborhood-level issues.

around Charles Ragin's innovative fs/QCA. For those unfamiliar with fs/QCA, the method allows for a form of inferential statistical analysis with as few as ten to twenty cases. But even if ten to twenty cases are a miniscule number from a traditional statistical perspective, they present an interesting challenge to the researcher who wants to use fs/QCA. The time demands of the traditional case study are simply too great to be compatible with Ragin's method. The third innovative feature of the study concerns the efficient, mixed-methods approach we devised to carry out the community fieldwork required to command our twenty cases. We will have much more to say about these various methodological innovations in Chapter 2.

Chapters 3 through 5 are organized around the principal research questions listed in the preceding text. We take up the first two of those questions in Chapter 3. Using simple descriptive measures, we seek to characterize the overall levels of emergent collective action across our twenty communities. We then turn our attention to the issue of variation in collective action. Are there factors – of our communities, their histories, or the specifics of the projects – that help us understand the variation in emergent action that we see in these locales? Chapter 4 takes up the all important issue of impact, or project outcome. Regardless of what explains variation in community collective action, does the level of that action appear to bear any relationship to the final outcome of the decision-making process? To put it as starkly as possible: does emergent opposition matter? Chapter 5 shifts the focus from our twenty communities to the nature of mobilized opposition in three different regions in the United States: the West Coast, the Gulf Coast, and the northeastern United States. For this chapter, we limit the analysis to the thirteen LNG projects that appear in our overall sample. We are interested here in exploring the distinctive form that opposition to LNG took in these three regions. In doing so, we are seeking to answer three questions: How did distinctive regional anti-LNG movements develop in two of our regions but not in the third? What mix of mechanisms shaped the emergence of regional anti-LNG movements on the West Coast and Gulf Coast? Why, given seemingly favorable circumstances, did opposition fail to "scale up" to the national level? That is, why did no national anti-LNG movement develop during this period? Finally, we bring the manuscript to a close in Chapter 6 by summarizing our main findings and, more importantly, teasing out the implications of the study for two audiences. The first of those audiences is simply the community of social movement scholars to whom this first chapter has largely been addressed. Our research, however, is really more a comparative analysis of local policy disputes than it is a study of social movements. As we will show in the Chapter 2, our cases are marked by the conspicuous absence of social movements. Heeding our own advice to break out of the narrow self-referential focus of most social movement scholarship, we also use Chapter 6 to speak to what we see as the important implications of the policy process literature for an understanding of our research.

2

Comparing Communities "At Risk" for Mobilization

Social movement scholars have long faced a methodological dilemma familiar to social scientists. We refer to the stark choice between two research methods that are, in their respective strengths and weaknesses, mirror images of one another. The choice is between "thin" large N studies that allow researchers to generalize to broader populations versus "thick" case studies that yield a rich, holistic understanding of the phenomenon in question, but without being able to tell us anything about the representativeness of the specific case on offer. These two options do not exhaust the methodological strategies available to movement researchers, but a broad scan of the extant empirical work on social movements would, we are convinced, show that these two approaches have been broadly modal in the field throughout the past thirty years.

The single case study has been especially popular with social movement scholars. A partial list of classic studies of single movements would include Adam (1987) on the gay and lesbian movement; Jenkins (1985) and Ganz (2009) on the farmworkers movement; Morris (1984), McAdam (1999 [1982], Luders (2010), and Andrews (2004) on the U.S. civil rights movement; Costain (1992), Rupp and Taylor (1987), Banaszak (1996), Evans (1979), and Mansbridge (1986) on various "chapters" or movements in the history of U.S. feminism; Amenta (2006) on the Townsend movement; Parsa (1989) and Arjomand (1988) on the Iranian Revolution; Meyer (1990) on the nuclear freeze movement; Smith (1991) on liberation theology; Schwartz (1976) on the Southern Farmers' Alliance; McVeigh (2009) on the resurgent Ku Klux Klan during the 1920s; Gould (2009) on ACT UP (AIDS Coalition to Unleash Power); and Walder (2009b) on the Beijing Red Guard movement. The fact that we could easily double or triple this list of citations only serves to make our point: it is probably fair to say that the case study has been the central arrow in the methodological quiver of social movement scholars.

It is not, however, as if movement analysts have been unaware of the "N of 1" problem, that is, the inability to generalize from a single case study.[1] This has motivated any number of researchers to invest in large N studies of various empirical phenomena related to social movements. To our knowledge there are no such studies of movements per se, but this hasn't stopped scholars from assembling large data sets of other related phenomena. By far the single largest group of such studies has taken the "protest event" as the fundamental unit of analysis. The list of movement scholars who have conducted some form of "protest event research" reads like a who's who of the field. The list would include published work by Tilly (Tilly, Tilly, and Tilly 1975; Tilly and Wood 2003); Tarrow (Tarrow 1989); McCarthy (McCarthy, McPhail, and Smith 1996); Olzak (Olzak, Beasley, and Olivier 2003); Soule (Soule 1997; Soule and Earl 2001; Soule and Olzak 2004); McAdam (McAdam 1999 [1982], 1983; McAdam et al. 2005; McAdam and Su 2002); Kriesi (Kriesi et al. 1995); Jenkins (Jenkins 1985; Jenkins, Jacobs, and Agnone 2003; Jenkins and Perrow 1977); and Koopmans (Koopmans 1993; 1995) among others. Interestingly, however, recourse to large protest event data sets doesn't necessarily solve the N of 1 problem. A number of these researchers use these numerically impressive data sets to study but a single movement, leaving the hoary issue of generalizability unaddressed. McAdam (1983, [1999] 1982), McAdam and Su (2002), Jenkins (1985), Jenkins, Jacobs, and Agnone (2003), Jenkins and Perrow (1977), Soule (1997), and Olzak, Beasley, and Olivier (2003) are but a few who employ the method in this way.

Other scholars have assembled large N data sets to study the dynamics of social movement organizations (SMOs) (Minkoff 1993, 1995, 1997), the role of organizational networks in social movements (Baldassare and Diani 2007; Diani 1995), city-level variation in riot activity (Eisenger 1973), differential recruitment to social movements (Klandermans and Oegema 1987; McAdam 1986, 1988; McAdam and Paulsen 1993; Passy 2001, 2003; Tindall 2004), and national-level variation in racial and ethnic conflict (Tsutsui 2004). Although greatly enriching our understanding of these aspects of contention, as a class of studies these works suffer from exactly the opposite problem of the traditional case study. In general, large N studies sacrifice depth of knowledge for inferential power. The researcher's knowledge of any particular case in his or her data set tends to be very "thin."

---

[1] Given the "inferential poverty" of the case-study approach, it is ironic that arguably the most influential theory in the field – political process – was based, at least initially, on research on the single case of the U.S. civil rights movement (McAdam [1999] [1982]; Morris 1984).

Given the stark trade-offs, you would have thought that more social movement scholars might have explored options designed to give them something of a methodological "middle ground" between the poles of large N research and the traditional case study. The problem is not so much a lack of interest on the part of scholars but simply the relative absence of true "middle ground" alternatives. Two examples will help make the case. The "paired comparison" has been used to great effect by scholars of contention – especially in political science – who still prize the richness of the case-study method but want to move beyond the obvious limitations of the single case (see, e.g., McAdam, Tarrow, and Tilly 2001). But for all of its virtues, it should be just as obvious that an N of 2 hardly solves the fundamental problem – generalizability, or lack thereof – inherent in the case-study approach.

At the other end of the spectrum, a few brave souls have sought to achieve a degree of inferential power while still amassing a fair amount of information on a relatively large number of cases. Perhaps the most celebrated of these attempts is the pioneering research reported in Gamson's 1975 book, *The Strategy of Social Protest*. Wanting to know something about the factors that made social movements more or less successful in their attempts to gain "new advantages" and "acceptance," Gamson and a number of grad students compiled a sample of fifty-three "challenging groups" drawn from the full sweep of U.S. history. Then using all the secondary information they could find on these groups, they coded them on a number of dimensions including organizational structure (e.g., centralized vs. decentralized), the nature of their goals (e.g., displacing vs. nondisplacing), their choice of tactics (e.g., violent vs. nonviolent), and so forth. The book reported a number of important findings and was among the most significant early works in the emerging field of social movement studies. From a methodological standpoint, however, there are two serious problems with Gamson's research design, which in the end has discouraged emulation by others. First, it is hard to empirically command as many as fifty-three cases, especially when you do no original research of your own. That means you are reliant on information gathered by others who invariably were not motivated by the same questions that are of interest to you. The second lacuna is the more important of the two. In taking the approach he did, Gamson was hoping to command sufficient statistical power to produce generalizable results. But in the end, the best he could do with his small sample size was to analyze the bivariate relationships between his independent and dependent variables. So he could, for instance, show – most controversially – that those groups that used violence were more success-ful in attaining "new advantages" than those that didn't, but he couldn't say anything about whether this relationship held net of other potential explanatory factors. For all the time and effort Gamson expended in compiling his sample and collecting and coding data on all fifty-three groups, in the end he lacked sufficient inferential power to produce the multivariate results required to answer his central research question more definitively. It is hard, however, to fault Gamson for the logic motivating his approach. He was, after all, seeking the very "middle ground" we aspire to here: a method of study that combined the depth of knowledge of individual cases with the inferential power of large N quantitative studies. In the end, it was the technical requirements of traditional regression analysis that undermined the approach rather than any inher-ent flaw in Gamson's logic.

MIDDLE GROUND APPROACHES

There have, however, been some encouraging efforts in recent years to fashion exactly the kind of middle ground approach to the evidentiary dilemma called for in the preceding text. Among these alternatives, we favor the comparative case analytic method pioneered by Charles Ragin (1987, 2000, 2008). But before we turn to a detailed discussion of this method, we would be remiss if we failed to acknowledge several other like-minded efforts. The most general, and arguably most important, of these efforts has been the tireless work of the likes of Henry Brady, David Collier, statistician David Freedman, and Charles Ragin, among others, to caution against naïve faith in quantitative methods, while exploring ways to deploy qualitative techniques in the service of causal inference (Brady and Collier 2004, 2010; Collier and Elam 2008; Freedman 2006, 2010; Ragin 2004). The touchstone publication in this sustained effort was Brady and Collier's 2004 book *Rethinking Social Inquiry*, which served as a rallying point and source of inspiration for those seeking a defensible third way between the extremes of mindless quantitative and nonsyste-matic qualitative research. Although eschewing any particular approach, the volume (and its worthy second edition) does more to stake out this methodological "middle ground" than any other publication to date.

The book also highlights several specific techniques available to those who seek to preserve the qualitative riches of the case study without giving up on the possibility of causal inference. Besides Ragin's method, the other well-established tool in this regard is the technique of process tracing (Bennett

2010; Drezner 1999; George 1979; George and Bennett 2005; George and Smoke 1974). At its most basic, process tracing involves the careful analysis of theoretically relevant pieces of evidence about a case that together are used to either support or disconfirm alternative causal explanations of the phenomenon in question. Although this basic description may suggest nothing so much as a lack of a distinctive method, specific "tests" have been devised over the years to aid the researcher in determining the necessity and sufficiency of the cause in question (Bennett 2010). When rigorously applied, the technique does appear to yield far more systematic and theory-driven results than your run-of-the-mill case study. Unlike Ragin's method, however, process tracing involves no formal (e.g., quantitative) technique for drawing causal inferences. Attracted by this latter feature of Ragin's approach, we chose to use the fuzzy set/Qualitative Comparative Analysis (fs/QCA) version of the "comparative case method" as the basis of our research.

CHARLES RAGIN AND THE EVOLUTION OF THE COMPARATIVE CASE METHOD

As a comparative historical sociologist, early on Ragin confronted the same dilemma that motivated the other middle-ground approaches sketched in the preceding text. He too wanted to study complex empirical phenomena – regime failure and ethnic political mobilization – related to the distribution of political power in society. Like Collier, George, and others, he sought a way to do so that would preserve the empirical integrity of each discrete case while still allowing for generalization to a broader population of cases. Ragin knew, however, that the numeric requirements of traditional quantitative analysis were prohibitively high; there was simply no way to reconcile the technical demands of the method with the desire to retain a depth of knowledge of each discrete case. So starting in the early 1980s, he began to develop a different approach to comparative case analysis based on set theory and Boolean algebra. The aim was to develop a comparative case method that would allow for a viable form of inferential analysis with something on the order of ten to twenty cases. He outlined the first version of his method in his 1987 book *The Comparative Method*. He followed that up with *Fuzzy-Set Social Science* in 2000 and *Redesigning Social Inquiry: Fuzzy Sets and Beyond* in 2008. Whereas the initial version required researchers to render all "causal" and "outcome conditions" (think of them as independent and dependent variables) in dichotomous terms, the new and improved fuzzy-set alternative – fs/QCA – allows analysts to define variables (or "conditions") as continuous fuzzy-set values ranging from 0 to 1 (e.g., 0.25, 0.40, and 0.65).

Besides allowing for meaningful inferential analysis with perhaps a tenth of the cases needed to employ traditional quantitative statistics, fs/QCA differs from the latter in at least two other crucial respects. First, instead of trying to maximize variation in one's independent and dependent variables, fs/QCA scores independent variables, or "causal conditions," and dependent variables, or "outcome conditions," in terms of membership in a set. Second, although traditional multivariate analysis seeks to measure simultaneously the strength of association between the dependent variable and each independent variable separately, the logic of fs/QCA is combinatorial. That is, Ragin's method expressly considers how causal conditions combine to create different pathways (or "recipes") to the same outcome. Because, in our view, emergent collective action is almost always produced by the confluence of several factors, the stress in fs/QCA on causal combinations makes the method ideal for our purposes. More importantly, it is the promise of the long sought-after "middle path" that recommends the method to us. But that still leaves unanswered the all-important question of what we conceive the empirical phenomenon of interest to be.

CASE SELECTION: COMMUNITIES "AT RISK" FOR MOBILIZATION

It should be clear from the critique we offered in Chapter 1 that we have strong views on the matter of what exactly scholars of contention should be studying. Here we simply restate what we argued in Chapter 1. For a good many research questions – in particular those that focus on internal movement dynamics, differential recruitment, or other such topics – it is perfectly appropriate to treat SMOs or other kinds of social movement actors as the units of analysis for research purposes. For those interested in understanding something about the factors and processes that shape the emergence and ultimate impact of social movements, the long-standing practice of selecting successful instances of mobilization for study must be regarded as a serious problem. "Selecting on the dependent variable" in this way cannot help but exaggerate the frequency of social movements, obscure the dynamics that likely shape their emergence, and overstate their general importance as forces of long-term social and political change.

The practical methodological solution to this problem is simple, if not necessarily easy to pull off. If we really want to understand emergent collective action we should be systematically comparing mobilization attempts or, better still, populations "at risk" for mobilization, rather than the rare and almost certainly atypical cases of mobilization that result in social movements. Only by looking at variation in the extent of mobilization across a broad range of cases can we ever hope to identify those factors and processes that mediate between latent and manifest contentious action. Although such studies are rare in the social movement literature, we were fortunate to be able to draw upon conceptual and methodological insights gleaned from three pioneering efforts published in the past fifteen years. The real innovators in this regard were Ed Walsh, Rex Warland, and D. C. Smith whose 1997 book *Don't Burn it Here: Grassroots Challenges to Trash Incinerators* summarized the results of their comparative study of eight communities that were located within three miles of proposed incinerator sites. Like us, they were interested in understanding the factors that shaped variation in community opposition and the ultimate fate of the proposed projects. Very much the same agenda motivated Daniel Sherman's exceptional comparative study of twenty-one communities designated as sites for the disposal of low-level radioactive waste. The results of Sherman's research were summarized in his 2011 book *Not Here, Not There, Not Anywhere: Politics, Social Movements, and the Disposal of Low-Level Radioactive Waste.* Finally, there is Daniel Aldrich's important book *Site Fights* (2008). Although also concerned with what he calls "divisive facilities," Aldrich's study differs from the other two – and the research reported here – in several important ways. For starters, Aldrich's study was conducted in Japan rather than the United States. The research also spanned a much longer period of time (1955–2005) and involved many more cases (nearly 500) than any of the other studies. Finally, Aldrich is not so much concerned with the dynamics of emergent opposition as he is with understanding why these particular communities rather than a matched sample of seemingly identical locales were targeted in the first place. More than anything, it is this focus on communities at risk that unites all of the studies – including our own – mentioned here.

From a methodological standpoint, however, shifting the phenomenon of interest in this way poses an interesting challenge. How does one study nonevents? Or more accurately, how do researchers identify or define communities at risk for mobilization? The answers to these questions are different for each of the aforementioned studies. In ours, however, federal environmental requirements allowed for simple and, we think, convincing answers to these questions. Energy infrastructure projects are almost always proposed by private companies. However, these proposals are subject to public review under the National Environmental Protection Policy Act (NEPA). NEPA requires the preparation and public dissemination of an Environmental Impact Statement (EIS) for all public and private projects when "a federal agency anticipates that an undertaking may significantly impact the environment" or is "environmentally controversial" (Environmental Protection Agency 2007).[2] We found the EIS process to be quite similar for all the projects in our study, although the particular agencies, individuals, and timelines involved differed.[3] As an example, we provide an overview of the process for onshore liquefied natural gas (LNG) projects.

As the designated lead agency, the Federal Energy Regulatory Commission (FERC), appointed by the president, is ultimately responsible for certifying EISs associated with onshore LNG proposals and subsequently approving or rejecting projects. Other agencies, such as the Coast Guard and the Army Corps of Engineers, assist with fact-finding and analysis, and outside consultants are often hired to complete the bulk of the data collection and analysis in the EIS with agency oversight. Nongovernmental groups and individuals may also participate at a minimum of two public hearings in which anyone can pose questions about the project and comment on the Draft EIS and Final EIS.

Outside of this formal process, the company may introduce the project and interact with the community in a variety of ways including by contacting the newspaper or holding press conferences, conducting company-sponsored informational meetings, meeting privately with elected officials and community leaders, and engaging with organizations that would potentially support or oppose the project. Supporters and opponents of a project are also known to create independent organizations, hold information sessions, gather and submit data on the costs and benefits of the project, contact the media in order to voice their opinions, and organize demonstrations.

---

[2] The term *controversial* has been interpreted by the courts to mean scientific disagreement over environmental impacts (Eccleston 2008).

[3] Although in our cases, the implementation of the EIS process, particularly related to public engagement, was quite similar across projects (the only major difference across projects was the number of public hearings), this might not be the case if we were to look at a wider range of decisions requiring EISs, particularly those unrelated to energy. Originally, we intended to use the level of public engagement as a variable in the analysis but found so little variation that it was dropped.

Because all large, potentially controversial, infrastructure projects are required to file an EIS, these records provide a population of communities at risk for mobilization. Thus our cases were drawn from the CSA Illumina's Digests of Environmental Impact Statements, which contains all EISs from the *Federal Register*. The population of cases was limited to proposals for new energy infrastructure projects in which a Final EIS was completed between 2004 and 2007 (N = 49). We chose projects that had completed a Final EIS in order to ensure that the window for potential collective action would be closed once the study commenced.[4] After drawing a random sample of twenty cases, three of them (15% of the sampling frame) were dropped because it was not feasible to conduct subsequent data collection due to the difficulty in attaining local newspaper records.[5] The sample was then supplemented with three California projects for which data had already been collected in previous research (Boudet 2010).[6] The final sample consists of twenty communities responding to eighteen projects[7] in twelve states regarding proposals for LNG terminals (13 cases), nuclear-related projects (2), a hydroelectric project, a wind farm, and a cogeneration project to supply electricity to an existing oil refinery.[8] Initial proposal dates span from early 2001 to 2005. Table 2.1 provides additional information about each

---

[4] Some of these cases stretched on for as many as 10 years so, although the completion of a Final EIS may indicate a lower level of community opposition (because projects experiencing high levels of opposition early on may be withdrawn before completion of the Final EIS), we felt strongly that we needed to know the ultimate fate of the sampled projects for the purposes of this research.

[5] For the purposes of efficiency, we searched newspapers using online databases and online archives provided by the publication. The publications in these communities had no obtainable electronic records making comparable levels of search logistically infeasible.

[6] Cabrillo Port was in the sampling frame. Mare Island and Long Beach were not due to the fact these projects were withdrawn before a Final EIS was completed.

[7] The number of communities does not correspond to the number of projects because some projects provoked reactions from two distinctly separate communities. For example, Ventura County and the city of Malibu, CA, both opposed the Cabrillo Port LNG proposal and Gloucester County, NJ, and New Castle County, DE, both responded to the Crown Landing LNG proposal. Gloucester County residents were relatively inactive when compared with New Castle County residents. Careful inspection of our cases will also reveal that there are only 17 unique communities. This is due to the fact that some communities received multiple proposals. Thus, when we discuss 20 communities, these can be thought of as communities by project. This is conceptually similar to the way time-series data presents state-years.

[8] Although we discussed limiting our analysis to LNG projects specifically because of the large number that were proposed during this time frame, we thought that it was important to expand to other types of energy proposals in order to allow for additional variation in project risk and political opportunity.

TABLE 2.1. *Case Information*

Community	Project	Type	Start	End
Aiken County, SC	Mixed Oxide Fuel Fabrication	Nuclear	2001	2005
Brazoria County, TX	Freeport LNG	LNG	2001	2004
Cameron Parish, LA	Sabine Pass	LNG	2003	2004
Cameron Parish, LA	Gulf Landing	LNG	2003	2007
Cameron Parish, LA	Creole Trail	LNG	2005	2006
Cassia County, ID	Cotterel Wind Power	Wind farm	2002	2007
Claiborne County, MS	Grand Gulf	Nuclear	2002	2007
Essex County, MA	Northeast Gateway	LNG	2004	2007
Gloucester County, NJ	Crown Landing	LNG	2003	2008
Long Beach, CA	Long Beach LNG Import	LNG	2003	2008
Malibu, CA	Cabrillo Port	LNG	2003	2007
Mobile County, AL	Compass Port	LNG	2004	2006
New Castle County, DE	Crown Landing	LNG	2003	2008
Providence County, RI	KeySpan	LNG	2003	2005
Riverside County, CA	Lake Elsinore Advanced Pumped Storage	Hydroelectric	2000	2007
San Patricio County, TX	Cheniere Corpus Christi	LNG	2003	2005
San Patricio County, TX	Vista del Sol	LNG	2003	2005
Solano County, CA	Mare Island Energy	LNG	2002	2003
Ventura County, CA	Cabrillo Port	LNG	2003	2007
Whatcom County, WA	Cherry Point	Cogeneration	2002	2004

of our cases. Figure 2.1 provides a map of project locations. The start dates noted in Table 2.1 correspond to either the first newspaper article on the project or the official filing of the project with the relevant federal agency, whichever came first. The end date in the table corresponds to the date of the proposal's rejection, withdrawal, or approval.

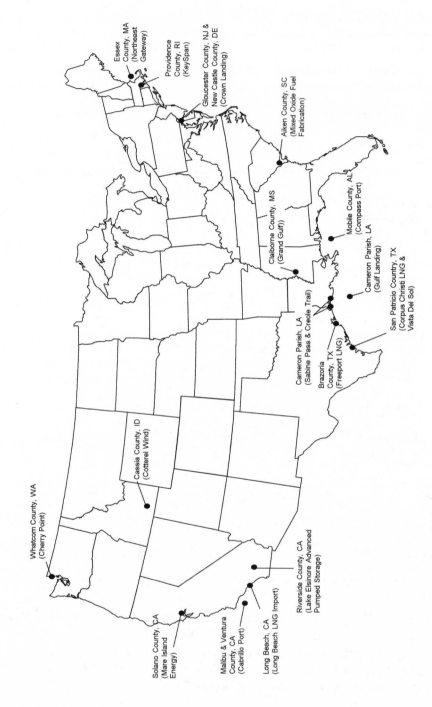

FIGURE 2.1. Map of selected cases

BUT HOW TO COMMAND TWENTY CASES? A NEW
APPROACH TO FIELDWORK

Eschewing the methodological conventions of social movement research, we imagined an ambitious and innovative comparative study of community response to the siting of eighteen environmentally risky energy projects affecting twenty locales. All well and good, but one major practical challenge remains. The method of fs/QCA may allow for meaningful analysis with as "few" as twenty cases, but empirically knowing twenty cases is no easy feat. The authors found this out when they collaborated with a number of other researchers on a study of variation in opposition to thirty water or pipeline projects proposed in the developing world between 1989 and 2005 (McAdam et al. 2010). Unable for practical and financial reasons to build a traditional fieldwork component into the project, we struggled – despite creative and Herculean efforts – to command all of our cases. The result was great variability in the amount and quality of the data that we had for any given case. What is more, the absence of sufficient data on some projects, made us drop those cases in favor of ones where more information was available. The problem is that amount of information may well be correlated with the dependent variable (e.g., level of project opposition). Projects that arouse more opposition are almost certain to generate more information, raising the serious specter of "selection bias" when it comes to those cases that ultimately made it into our sample.

Our experience with this otherwise groundbreaking study convinced us of the need to build a serious fieldwork component into the current project. But as experienced field-workers, we knew that there were perils aplenty down this road as well. The biggest problem had to do with the temporal and financial demands of the traditional fieldwork method. Relying exclusively on the method, it had taken the second author nearly three years to complete case studies of opposition to two proposed LNG projects in California. There was simply no way to reconcile our target of twenty cases with the amount of time, energy, and financial resources that Boudet had expended on each of her two cases. We were once again looking for a "middle ground," this time between the excessive demands of traditional fieldwork and the efficient but ultimately "thin" sources of data we had employed in the cross-national study of water and pipeline projects. To the two methodological innovations discussed in the preceding text – fs/QCA and the shift in focus from the social movement to a community at risk for mobilization – we now add a third. We refer to it as advanced preparation fieldwork (APF).

APF involved two substantial online data-collection efforts per case that, in turn, preceded and served as the basis for an intensive seven-to-ten-day traditional fieldwork visit to the community in question. The visit was given over primarily to interviews and various forms of archival research. What follows is a brief description of each of these three components of the APF approach.

1. The Community Profile

We began our research on each of our cases by using a wealth of online sources to compile a detailed social, economic and political profile of the community in question. This online data-collection effort was designed not only to give us a detailed "snapshot" of each of our twenty communities but also to allow us to compare the communities to each other. We collected information on many different community factors (for a complete list, see Appendix A). However, for the final analyses reported here, we zeroed in on the following factors that we felt were most relevant for mobilization:

a. Demographic Factors
Population, median income, unemployment, percent with a college degree, median home value, and home-ownership from the 2000 Decennial Census and the American Communities Survey (ACS);

b. Political Factors
Voter turnout as the percentage of registered voters who voted in the most recent presidential election prior to the proposal announcement from county clerk or state Web sites, and

c. Organizational Capacity
Nonprofits per 1,000 people from the National Center for Charitable Statistics 2000 Business Master File.

Our ability to assign meaningful "fuzzy-set" values to these communities on a number of key independent variables (or causal conditions) depended critically, as we will explain in the following text, on the resulting comparative portrait of our twenty research locales. (See Appendix B for raw values for each of these community factors by case.)

2. "Before" and "After" Newspaper Analysis

The online profile afforded us a general static snapshot of each of our communities. The second component of APF was designed to give us a more dynamic understanding of our twenty research sites. The goal was to identify the newspaper of record for each of our cases and to systematically read and code information from the paper for (a) a year prior to the announcement of the proposed project and (b) the duration of the decision-making process concerning the project. This second phase of the newspaper analysis/coding lasted anywhere from a minimum of one year to a maximum of seven years. Table 2.2 provides information about newspaper data we collected on each case. The goal of the newspaper analysis was slightly different in the "before" and "after" phases. In the "before" phase, we were focused on answering four primary questions: What were the major issues or conflicts in the community on the eve of the project announcement? Did these issues or conflicts involve any controversial land-use questions that might have helped set the interpretive frame for the proposed project? Who were the main "players" in town on the eve of the announcement? What was the general editorial stance of the newspaper at the time of the announcement?

The goal of the "after" coding was nothing less than the compilation of a comprehensive history of the decision-making process with regard to the proposed project. Toward that end, we kept a log of all organizations that sought to involve themselves in the community response to the project. Besides simply listing the organization, we also kept a running tally of how many times the organization was mentioned in articles on the project. We also noted the organization's stance (for, against, or neutral) toward the project and whether that stance changed over the course of the decision-making process. We kept a similar log of all individuals who likewise sought to involve themselves in the community debate over the proposed project. Many of these individuals were formally affiliated with one of the organizations included in the log described in the preceding text, and where that was the case, the association was noted. Typically, however, there were also some number of unaffiliated individuals who sought to shape the community response to the proposed project, and they were noted as well. We generated a running chronology of events that spanned the life of the decision-making process concerning the project. These events ranged from, among others, "community meetings," to court or other governmental actions impacting the project, to the rare contentious event organized by opponents or proponents of the project. Besides noting its date, the events were also coded for "stance" (for, against, or neutral) and, where relevant, for the number of people taking part.[9] Finally, we also

[9] We triangulated from interviews and EIS comments in order to verify information reported in newspapers.

TABLE 2.2. *Newspaper and Interview Data Collection by Case*

Community	Newspaper(s)	Articles (#)	Letters and Editorials (#)	Interviews (#)	Average Interview Time (hours)
Aiken County, SC	Augusta Chronicle	120	53	6	1.48
Brazoria County, TX	Brazosport Facts	75	12	8	1.07
Cameron Parish, LA (Gulf Landing)	Cameron Parish Pilot The Times-Picayune (New Orleans)	92	0	6	1.13
Cameron Parish, LA (Sabine Pass)	Cameron Parish Pilot Beaumont Enterprise	79	1	6	0.63
Cameron Parish, LA (Creole Trail)	Cameron Parish Pilot	70	0	6	0.63
Cassia County, ID	Twin Falls Times-News	60	14	9	1.38
Claiborne County, MS	Vicksburg Post	75	9	13	1.10
Essex County, MA	Gloucester Daily Times	144	46	4	0.77
Gloucester County, NJ	Gloucester County Times	110	28	3	0.58
Long Beach, CA	Long Beach Press-Telegram	154	101	8	1.62
Malibu, CA	Malibu Times	89	122	14	1.30
Mobile County, AL	Mobile Press-Register	66	71	9	0.95
New Castle County, DE	Wilmington News Journal	130	12	8	0.84
Providence County, RI	Providence Journal	108	40	6	1.27
Riverside County, CA	Riverside Press-Enterprise	88	7	6	1.27
San Patricio County, TX	Corpus Christi Caller-Times	64	6	8	0.77
San Patricio, TX (Cheniere) (Vista del Sol)	Corpus Christi Caller-Times	64	4	8	0.77
Solano County, CA	Vallejo Times-Herald	81	237	16	1.13
Ventura County, CA	Ventura County Star	139	417	20	1.30
Whatcom County, WA	Bellingham Herald	16	5	7	0.88
Total		1,620	1,185	172	–
Average/Case		81	59.3	8.6	1.0

generated a running, chronological log of all editorials or letters printed in the paper regarding the project, noting the author's stance on the enterprise. Taken together, the newspaper data afforded us a more dynamic sense of the community on the eve of the project announcement as well as a comprehensive history of the official decision-making process and the informal community debates and conflicts (if any) that unfolded over the course of the review process.

3. Intensive Fieldwork

The data collected as part of the first two components of APF typically filled at least one and sometimes two very large binders. These became the case files for each of the communities being studied. Informed by the relevant case file, one of the three researchers on the project would then schedule a short (typically 5 to 10 days) visit to the community in question. Although not precluding other forms of research (especially archival), the centerpiece of the fieldwork trip was interviews conducted with key informants gleaned through the newspaper analysis. The number of interviews ranged from a low of four to a high of twenty-four (see Table 2.2). Appendix C provides a complete listing of all interviews conducted. In a typical case, we would seek to interview one or more informants drawn from the following categories of participants: officials centrally involved in the decision-making process, project spokespersons, opposition figures, local proponents of the project, and a reporter who covered the case. Wherever possible, we also tried to interview someone with knowledge of the case but who remained neutral throughout the process. The aim of the fieldwork component was simply to confirm or refine the understanding of the case that we had already developed through the analysis and coding of the newspaper data.

FROM CASE FILES TO FUZZY-SET VALUES

In most cases, we found that our understanding of the evolving community response to the proposed project and the official decision-making process was largely affirmed by the fieldwork phase of the research. Still the interviews added crucial nuance and details that hadn't come through in the newspaper analysis and coding. Armed with the mass of information collected in all three phases of the project, we were now in a position to translate our knowledge and understanding of the case and the community into a form amenable to analysis using the fs/QCA method. In practical terms, this meant assigning fuzzy-set values to the key "causal conditions" and "outcome conditions" around which the analysis was to be structured. What follows is a discussion of all of the items in each of the two categories.

Coding Fuzzy-Set Membership

We transformed our raw data to fuzzy-set values in two different ways. For the majority of conditions, we coded in simple binary terms or assigned fuzzy-set membership to communities based on the percentile value for each condition relative to other communities. For example, one of our causal conditions is "civic capacity." A community in the 40th percentile for civic capacity was assigned a fuzzy-set value of 0.4 placing it more out of than in the set of communities with civic capacity. For more complicated concepts like risk, political opportunity, and civic capacity we often combined measures to create final, multidimensional versions of the variables/causal conditions. Let's return to the example of civic capacity. We used three items of information – the number of nonprofits per capita, voter turnout, and the percentage of adults with at least a bachelor's degree – to assign a community to the set for civic capacity. In each case, we assigned fuzzy-set membership scores for each of these three components (e.g., number of nonprofits per capita individually) to normalize the values, added them together, and then reassigned fuzzy values based on the total. For example, the number of nonprofits per capita in Whatcom County fell in the 80th percentile relative to the other nineteen communities and so was assigned a 0.8. The voter turnout rate fell in the 99th percentile and was assigned a 1. The percentage of college-educated citizens fell in the 60th percentile and was assigned a 0.6. Next, these three normalized values were summed together giving Whatcom County a total of 2.4 (0.8 + 1 + 0.6) for civic capacity. The summed scores for all twenty communities ranged from 0.2 to 2.8. With a 2.4, Whatcom County fell in the 80th percentile and was assigned an overall score of 0.8 putting it mostly but not fully in the set of communities with high civic capacity. After we assigned communities fuzzy-set values for each condition, we double-checked the values with our expert knowledge of each case derived from the newspaper research and field interviews. This confirmed the assigned score and that we were using appropriate proxies for our concepts.

For two concepts (project outcome and previous oppositional experience), this type of coding method was not feasible given the more qualitative nature of the raw data collected. For these conditions, we transformed our expert knowledge of the cases into a coding scheme. We conducted sensitivity analyses on the two conditions coded in this manner to ensure that the results of our analysis were not an artifact of our coding scheme but were related to the concept at hand.

Appendix B provides more detailed explanations of each of our causal and outcome conditions, values for each data point, corresponding scores for each data point, and total scores for each condition. Table 2.3 provides the fuzzy-set scores for all outcome and causal conditions by case.

OUTCOME CONDITIONS

The study focuses on two outcomes of interest: level of oppositional mobilization and the ultimate fate of the project. We take up each of these two key dependent variables or "outcome conditions" in turn.

Oppositional Mobilization

In deciding whether a community belongs to the set of communities that mobilized, we incorporate three separate scores reflecting the use of noninstitutional tactics, evidence of coordinated action, and the overall volume of responsive action in the community.

Noninstitutionalized forms of action have long been regarded as a hallmark of social movements as compared to other collective change vehicles. Consider the following typical definition of a social movement: "organized efforts, on the part of excluded groups, to promote or resist changes in the structure of society that involve recourse to *noninstitutional forms of political participation*" (McAdam 1999 [1982]: 25; emphasis added). What exactly do we mean by noninstitutional forms of action? We regard forms of action that take place outside the institutionalized systems of collective claims making and conflict resolution as noninstitutional. Communities that featured nonroutine or noninstitutional tactics were in the set on this dimension (scoring 1) and out otherwise (scoring 0). But noninstitutional actions hardly exhaust the responses open to citizens in such cases. In addition to noninstitutional tactics, we also looked for evidence of collective behavior in institutionalized settings such as lawsuits, coordinated appearances at EIS meetings, and community-

TABLE 2.3. *Fuzzy-Set Scores on All Causal and Outcome Conditions by Case*

Community	Risk	Political Opportunity	Civic Capacity	Similar Industry	Previous Oppositional Experience	Economic Hardship	Oppositional Mobilization
			Causal Conditions				Outcome Condition
Aiken County, SC	0.2	0	0.4	1	0	0.2	0.6
Brazoria County, TX	0.6	0.6	0.2	1	0.6	0	0.2
Cameron Parish, LA (Gulf Landing)	0	0.6	0	1	0	0.2	0.6
Cameron Parish, LA (Creole Trail)	0.2	0	0.2	1	0	0.2	0
Cameron Parish, LA (Sabine Pass)	0	0	0	1	0	0.2	0
Cassia County, ID	0	0	0.4	0	0	0.6	0.2
Claiborne County, MS	0	0	0.2	1	1	1	0.8
Essex County, MA	0.4	0.6	0.6	0	0.8	0	0.2
Gloucester County, NJ	0.6	0	0.4	1	1	0.2	0.2
Long Beach, CA	0.8	0.8	0.4	1	0.6	0.6	0.6
Malibu, CA	0.2	0.6	0.8	0	0.6	0	0.8
Mobile County, AL	0.4	0.8	0.6	1	1	0.8	0.6
New Castle County, DE	0.6	0.4	0.8	0	1	0.2	0.6
Providence County, RI	1	0	0.6	1	0.6	0.2	0.2
Riverside County, CA	0.6	0.8	0.2	0	0.6	0.6	0.6
San Patricio County, TX (Cheniere)	0.4	0	0	1	0	0.6	0
San Patricio County, TX (Vista del Sol)	0.4	0	0	1	0	0.6	0
Solano County, CA	0.6	1	0.6	0	0	0	0.8
Ventura County, CA	0.2	0.6	1	0	1	0	1
Whatcom County	0.6	0.6	0.8	1	0.8	0.2	0

initiated information and strategy meetings. Communities that exhibited any of these types of collective activities were in the set on this dimension (scoring 1) and out otherwise (scoring 0). Finally, we scored communities based on whether they belong to the set of communities that experienced high levels of contention in response to these projects. To calculate this score we used information about the number of collective activities mentioned in the preceding text, as well as information about activities that can express opposition by single individuals. This includes the number of letters to the editor, the number of people who commented during review hearings, and attendance at protest events, community events, and so forth. Table 2.4 shows raw data on mobilization, scores on each dimension, and the final mobilization score assigned to each community. To be in the set of communities that mobilized (score higher than 0.5, the point of maximum ambiguity), the community had to show evidence of collective and noninstitutional behavior. Under these conditions the community then received a mobilization score equal to its volume score if the volume score was 1 or 0.8; the case received a score of 0.6 if the volume score was equal to or less than 0.6. Communities with evidence of collective behavior or noninstitutional behavior (but not both) fell out of the set and received a score equivalent to their volume score. Communities with no evidence of collective behavior or noninstitutional action scored a zero.

Project Outcome

We were not only interested in mobilization but also in the eventual outcome of the project. Was the project approved? Was it built? Was it withdrawn by proponents? Was it rejected by state or federal officials? This information is important in order to determine the effectiveness of opposition efforts. Chapter 4 provides more information about the relationship between mobilization (or the lack thereof) and project outcome.

CAUSAL CONDITIONS

We considered a host of individual variables, or causal conditions, before focusing on the following six key measures.

Risk

The siting of an industrial facility presents the possibility of many different objective "threats" (e.g., to public safety, health, environment, and quality

TABLE 2.4. *Fuzzy-Set Membership in the Set of Mobilized Communities*

Location	Letters	Speakers	Coordinated Appearances	Public Meetings	Protest Events	Lawsuit (y/n)	Noninstitutional	Collective	Volume	Final Score
Ventura County, CA	246	138	5	3	8	n	1	1	1	1
Claiborne County, MS	2	9	1	2	1	y	1	1	0.8	0.8
Solano County, CA	192	120	1	3	4	n	1	1	0.8	0.8
Malibu, CA	92	138	6	3	5	n	1	1	0.8	0.8
Aiken County, SC	12	21	0	1	2	y	1	1	0.6	0.6
Mobile County, AL	47	38	0	1	1	n	1	1	0.4	0.6
New Castle County, DE	9	20	1	1	1	n	1	1	0.6	0.6
Cameron Parish, LA (Gulf Landing)	0	1	2	0	2	y	1	1	0.6	0.6
City of Long Beach	65	22	5	1	2	n	1	1	0.6	0.6
Riverside County	5	72	5	1	1	n	1	1	0.6	0.6
Essex County, MA	19	26	0	1	0	n	0	1	0.2	0.2
Gloucester County, NJ	6	13	0	1	0	n	0	1	0.2	0.2
Providence County, RI	15	17	0	1	0	n	0	1	0.2	0.2
Brazoria County, TX	5	7	1	0	0	n	0	1	0.2	0.2
Cassia County, ID	5	5	1	1	0	n	0	1	0.2	0.2
Cameron Parish, LA (Creole Trail)	0	1	0	0	0	n	0	0	0	0
Cameron Parish, LA (Sabine Pass)	0	1	0	0	0	n	0	0	0	0
San Patricio County, TX (Vista del Sol)	2	0	0	0	0	n	0	0	0	0
San Patricio County, TX (Cheniere)	0	0	0	0	0	n	0	0	0	0
Whatcom County, WA	2	4	0	0	0	n	0	0	0	0

of life) around which a community can mobilize.[10] In the technical and planning literature, these aspects of a facility are referred to as risks (Boholm 2004; Freudenburg 2004; Schively 2007). To score this condition, we tried to capture measures of common community concerns associated with large projects, including risk perceptions of the project by type, potential safety risks by proximity, and property values. In many ways, this condition incorporates the "classic" view of factors that elicit a NIMBY (Not in My Backyard) response (Dear 1992; Gallagher, Ferreira, and Convery 2008; Hunter and Leyden 1995; Lober 1995; Vajjhala and Fischbeck 2006).

The condition is coded so that a community's value corresponds to the average level of threat it experiences across all three areas (project risk, safety, and quality of life) in the same way as described on page 44 for civic capacity. To score project risk, we ranked the projects by type according to findings by Slovic (1987) about which types of projects people perceive to be the most risky. Our two nuclear-related projects received the highest scores. The Mixed Oxide fuel fabrication plant, which will repurpose weapon-grade plutonium for use as nuclear reactor fuel, scored a 1. The Grand Gulf nuclear power plant received a score 0.8. Natural-gas–related projects – including offshore, open-loop LNG, onshore LNG, and cogeneration projects – received scores of 0.6. The term *open loop* refers to the system of regasifying the LNG. Open-loop systems use sea-water to regasify LNG, while closed-loop systems use either a portion of the imported gas or ambient air. The use of open-loop regasification has been shown to pose a significant risk to fisheries relative to other LNG technologies. Offshore, closed-loop facilities received scores of 0.4. Our hydroelectric project scored 0.4, and the wind project scored 0.

Safety threat was determined relative to the other communities by combining information on population density and the proximity of the project to the community. Similarly, property threat is calculated relative to other communities taking a combination of the proximity of the project to the community and the median home value. This reflects the average

---

[10] We admit that setting objective measures of threat for inherently ambiguous technologies and policies is tricky. Here we attempt only to capture affects for which citizens can be certain and that would be likely to affect locals equally. We also ran the analysis with only project type, which is a more straightforward measure of threat derived from survey work conducted by risk analysts, and got almost identical results. We report objective measures because our point is that despite facing similar proposals communities react in diverse ways. Thus using only project type seems to stack the deck in favor of our hypothesis that it is contextual conditions that shape the perception of threat that matter, not the inherent characteristics that are broadly similar across cases.

community member's attachment to their home and the impact a potential project could have on personal finances.

Political Opportunity

In recent years, political opportunity has been criticized for subsuming many conceptually different ideas, mechanisms, and measurements (Gamson and Meyer 1996; Goodwin and Jasper 1999; Meyer and Minkoff 2004). Here we limit our definition of the concept to refer only to how open the institutionalized political system is to the claims of movement actors.

To score political opportunity, we used the same three indicators as Boudet (2010): electoral vulnerability of decision makers, temporal proximity to an upcoming election, and jurisdiction. First, following Skocpol (1985), we note that bureaucrats are able to act independently of public pressure. In contrast, elected officials are much more susceptible to public opinion. Therefore decision-making bodies that are made up of elected officials will present opponents with more political leverage than those that are highly bureaucratic. Thus we measure the proportion of all decision makers that were elected officials and score communities relative to one another based on this proportion. Second, because elected officials are held accountable through the election process, an election presents another window of political opportunity. Thus communities in which elected decision makers were up for reelection during the review process were in the set on this indicator (scoring 1) and out otherwise (scoring 0). Finally, studies show that local actors have less influence over decisions under national jurisdiction (Andrews 2004; Boudet and Ortolano 2010). Thus we argue that there is more political opportunity when local officials are responsible for the ultimate decision compared to when state or national officials are primarily responsible. We scored the jurisdiction of decision makers as follows: a case received 1 if the elected officials were local, 0.6 if there was at least one local elected official, and 0 if there were no local elected officials on the decision-making body. These three aspects were then combined into a single score of membership in the set of opportune political structures as described in the preceding text.

Civic Capacity

As described in the preceding text, three proxies were used to assign communities to the set with relatively high levels of civic capacity.

Following Molotch, Freudenburg, and Paulsen (2000), Sherman (2011), and Boudet (2010), the number of nonprofits per capita was used as a proxy for organizational capacity. Following Hamilton (1993), voter turnout figures were used as a proxy for community involvement in the democratic decision-making process. Finally, the percentage of community members with a college education was used as a proxy for community members' general knowledge and sense of efficacy. Studies show that education levels are not only a major predictor of volunteerism and community involvement (Musick and Wilson 2008) but also are the only consistent determining factor of oppositional attitudes to locally unwanted land uses (Freudenburg and Gramling 1993).

Similar Industry

We use the presence of a similar industry in the community as a proxy for a community's familiarity and comfort with a technology similar to the one proposed. This condition was scored based on the research of each individual case. Communities were assigned 1 if the exact industry was already operating at the time of the proposal and 0 otherwise. The presence of similar industry may denote the type of "developmental channelization" described in Gramling and Freudenburg (1996, 483). They argue that, "As the various components of the human environment become adapted to a given form of development activity, there is a tendency for new skills, knowledge, tools, networks, and other resources to be built up around that activity . . . . [This process] narrows a region's options, because time, resources, human capital, are devoted to a particular developmental scenario, sometimes limiting the options for alternative scenarios" (Gramling and Freudenburg 2006, 456). We add to this argument the idea that the presence of a similar industry, particularly one that has operated without incident, creates a certain level of comfort and acceptance of the potential risks associated with this industry when weighed against immediate benefits of jobs and economic gain. Thus these communities are more likely to accept future proposals for projects of a similar nature.

Previous Oppositional Experience

Previous experience opposing such a project or a similar issue may sensitize a community to a newly proposed project. This condition was coded to reflect the existence, recency, and similarity of such an experience. Communities were assigned a score of 1 if they opposed the exact same type of project five or fewer years prior to the proposal. They scored a 0.8 if they had a previous oppositional experience with a similar project any time in the past. They scored a 0.6 if they experienced a major dispute regarding land use (but not specifically energy) in recent years. Otherwise, the community scored 0.

Economic Hardship

We suggest that the potential economic benefits of a project may cause a community to downplay potential drawbacks and limit the perception of threat. Because the benefits of a project to a local community are typically economic, this desire should stem from a state of economic hardship such that the community would actively support investment. Each community was scored relative to the others based on its unemployment rate reflecting a need for jobs and its relatively low-median income reflecting need for general investment. The final score is determined using the method described in the preceding text.

CONCLUSION

It is conventional to include a methods chapter in a research monograph, but in this case we see the chapter as more central to the overall aims of the book than it normally might be. Besides our central theoretical goal of redressing the movement-centric bias in the field by studying communities at risk for mobilization rather than movements per se, from the very beginning we conceived of the project as an attempt to develop an alternative to the methodological conventions of social movement research. Equally dissatisfied with "thin," large N studies of protest events and rich but nongeneralizable case studies of this or that movement, we sought a middle ground between these two modal "poles" of social movement scholarship. The alternative on offer here combines at least some of the richness of the traditional case-study approach with the novel inferential possibilities of fs/QCA. But to take advantage of these "inferential possibilities," Ragin's comparative case approach still requires the researcher to have empirical command of at least fifteen to twenty cases. The problem is it would be prohibitively expensive and time consuming to use traditional fieldwork methods to "know" this many cases. Our response to this dilemma is what we have termed APF. By using readily available online sources to compile a detailed "before" profile of each of our twenty communities and a systematic analysis of newspaper data from the formal announcement of the project through its final resolution, we were able to greatly compress the amount of time needed for traditional fieldwork in each of our twenty locales. Armed with one to two thick binders of information on the case, the field-worker was able to identify the most promising subjects to interview in advance of his or her visit to the community, arrange those interviews, and most importantly, devise an interview schedule based on a rich understanding of the case and community at hand. In all but a few instances, this enabled us to limit the amount of fieldwork needed to refine our understanding of the case to a seven-to-ten-day period. The resulting efficiencies allowed us to command our twenty cases with little difficulty. Thus APF is the perfect complement, in our view, to the comparative case method proposed by Ragin. Having spelled out what we see as the important methodological contributions of the research, we are now ready to turn to the actual analyses of our data.

3

Explaining Variation in the Level of Opposition to Energy Projects

In Chapter 1, we spelled out four principal research questions that we hoped to address in the book: How much emergent collective action do we see across the twenty communities we are studying? What causal conditions explain variation in the level of mobilization in these communities? Net of other factors, what influence, if any, does the level of mobilized opposition have on the outcome of the proposed project? Finally, why did opposition to one kind of energy project – liquefied natural gas (LNG) terminals – grow into broader regional movements in some parts of the country but not others? In this chapter, we take up the first two of these questions. We aim to establish something of a baseline of just how much emergent collective action we see across the twenty communities we are studying and then move to see if we can identify certain mixes of factors – or "recipes" in the language of fuzzy set/Qualitative Comparative Analysis (fs/QCA) – that help account for variation in level of mobilization across these locales. We begin with a basic description of the level and forms of emergent collective action we found in the twenty communities that we have come to know so well throughout the past few years.

ESTABLISHING A BASELINE

A newcomer to the United States would be forgiven if, in reading the social movement literature, she or he imagined that our communities were awash in protest activity. In a 2005 article, McAdam and colleagues (2005: 2) argued that the field's initial engagement with and continued interest in the movements of the 1960s had "created a stylized image of movements that threatens to distort our understanding of popular contention ... in the contemporary U.S. ... This stylized view tends to equate movements with:

- Disruptive protest in public settings
- Loosely coordinated national struggles over political issues
- Urban and/or campus-based protest activities
- Claim making by disadvantaged minorities."

The authors went on, in the same article, to present over-time data on "protest events" in Chicago that showed that, even at the peak of the New Left "protest cycle" in 1970, movement activity did not conform to the stylized view sketched in the preceding text. In that year, protests in the Windy City were overwhelmingly oriented to local issues and only rarely (less than 5% of the time) featured the kinds of disruptive elements – for example, arrests, violence, and property damage – that we tend to associate with the movements of the period. The disconnect between the stylized image and the empirical reality has only grown greater with time. The implication is clear: by focusing disproportionate attention on the most contentious struggles of the 1960s and 1970s (and their contemporary heirs), movement scholars have encouraged a view that exaggerates the frequency and character of popular protest activity. The overwhelming methodological injunction to select successful instances of mobilization rather than some other empirical phenomenon for study – for example, mobilization attempts or, as in our case, communities at risk for mobilization – only reinforces the tendency to see movements as more frequent and more intensely disruptive than we believe to be the case.

Still, notwithstanding our skeptical view of the matter, we know very little about what the "baseline" of local movement activity looks like. Even as the study of social movements has expanded exponentially throughout the past thirty to forty years, to our knowledge, no one has sought to generate any empirical estimate of any such baseline. Nor is that exactly what we have tried to do with our research. By choosing to focus on locales that have been designated as sites for large, environmentally risky energy projects, we too would seem to have stacked the deck in favor of higher levels of emergent collective action than might be expected at random. Still ours is a far cry from studies of this or that mature movement, protest cycle, or social movement organization. Most importantly, in choosing sites for our research we have selected twenty communities at random rather than "cherry picking" cases that we know featured substantial grassroots opposition to the proposed projects. Thus although we can't claim that our data represent something approaching a true baseline of local collective action, we are at least edging in that direction. More specifically, the question we can answer is: how much emergent collective action do we see across a random sample of locales that we might think of as communities at risk for mobilization by virtue of their designation as sites for "risky" energy projects?

One might object to our characterization of these communities as ripe for mobilization. Isn't it possible, reflecting the central premise of the environmental justice movement, that these communities were selected as the sites for the proposed projects precisely because they were seen as incapable of collective action? Setting aside the fact that the empirical evidence for such selective targeting of polluting facilities is decidedly mixed (Adeola 2000; Grant, Grant, and Bergesen 2002; Muthukumara and Wheeler 1998; Simon 2000), we think our cases are much less subject to selective siting than other environmentally risky facilities. Consider the case of LNG terminals that make up nearly two-thirds of our proposed projects. Unlike most environmentally risky technologies, LNG terminals have exceedingly restrictive site requirements. Three requirements are worth noting. The most obvious is that LNG facilities must be located on or near the coast. Still, given that the continental United States boasts nearly six thousand miles of coastline, in and of itself, this requirement is not all that restrictive. It's the other two requirements that dramatically constrain which coastal sites are suitable for the technology. Besides coastal location, LNG terminals require deep-water channels and close proximity to existing pipeline infrastructure. Together these two site requirements dramatically restrict the coastal locations that match the demands of the technology. One other comment is in order here. Had the companies proposing these projects been overly concerned with avoiding communities that were likely to oppose the projects, it is highly unlikely that they would have slated so many LNG projects for the environmental hotbed of the West Coast.

A FINAL METHODOLOGICAL PRELIMINARY

The protest event has emerged as the preferred operational measure of social movement activity within the field. But although it is a convenient metric, there are real problems with equating contentious collective action with protest events. For starters, the protest event is a decidedly nonrepresentative form of movement activity. Although definitions of social movements vary, most are similar to the one that we offered in Chapter 2: "A social movement is a loosely organized, sustained effort to promote or resist change in society that relies at least in part on noninstitutionalized forms of collective action." The stress on noninstitutionalized forms of collective action is necessary in order to distinguish movements from other collective change efforts – such as public interest lobbies, electoral campaigns, or public education initiatives – that rely exclusively or overwhelmingly on "proper channels" for their effectiveness.

The empirical stress on protest events, however, has had the unfortunate effect of focusing attention on these noninstitutionalized forms of action at the expense of the more routine activities that tend, in most cases, to make up the vast majority of what movements actually do. Relative to mass mailings, press releases, educational campaigns, and fund-raisers, true protest events are rare occurrences. This is true even for movements that we have come to see as synonymous with extreme forms of protest behavior. For every celebrated sit-in or protest march, U.S. civil rights activists deployed a much broader and far more conventional array of tactics than the canonical view of the movement suggests. Moreover, we have no doubt that over the course of the entire struggle civil rights forces relied far more on routine forms of action than the rare instances of disruptive protest for which the movement is duly famous. For the vast majority of movements, the balance between noninstitutionalized and institutionalized forms of action is shaded even more dramatically in the direction of the former. The only exceptions to this rule would seem to be actual revolutions. But having made protest events the empirical coin of the realm, movement analysts tend to distort this balance and fail to measure the more routine forms of collective action that are typically the hallmark of contentious politics.

This is all by way of saying that we aspired, from the very beginning of the project, to a much broader and more fine-grained measurement of contention in response to a proposed project. More specifically, we sought to differentiate between and empirically measure six different forms of emergent action in support of or in opposition to the projects. These six forms of action include (1) letters to the editor regarding the project, (2) speakers who appeared at the mandated Environmental Impact Statement (EIS) hearings on the project, (3) coordinated appearances by supporters or opponents of the project at public events (other than EIS hearings) linked to the project (e.g., city council meetings or informational meetings sponsored by the lead company on the project), (4) public meetings organized by opponents or proponents of the project, (5) lawsuits brought by opponents or proponents, and (6) protest events initiated by opponents or proponents. Note that the first two forms of behavior are individual rather than collective in nature.[1] Although we sought to monitor and measure these forms of individual behavior carefully, they did not figure in our operational definition of emergent contention, or *opposition mobilization*, as we have termed the outcome condition of interest. To be judged an example of the latter, the action had to satisfy two criteria. It had to be coded as *noninstitutional* and *collective*. All of the other remaining forms of action listed in the preceding text – that is, forms three through six – are collective in nature, satisfying the second of our two criteria. But only one of these forms of action – protest events – was deemed as being noninstitutional in nature. Only those communities that featured forms of action that satisfied both of these criteria were seen as having experienced emergent collective action or opposition mobilization in response to the proposed project. Using these various definitions, we are now in a position to describe the level of various forms of action we saw in our communities in response to the proposed projects.

SO HOW MUCH COLLECTIVE ACTION DID WE FIND?

The simple answer to the preceding question is: not much.[2] Table 3.1 reports, for the six categories described in the preceding text, the number of actions recorded for each of our twenty cases.

By any standard, perhaps none so much as the expectations set by social movement scholarship, the levels of action reported in the table seem paltry. The modest amount of activity is fully consistent with the exceedingly low levels of "protest" behavior reported by Verba, Schlozman, and Brady in their influential 1995 survey of American civic life. Even the levels of individual action in response to the projects seem, with a few exceptions, to be modest. The median number of letters and speakers across all cases is just 5.5 and 15, respectively. But even these numbers seem impressive when compared to the more collective forms of action reported in the table. The median number is 1 or less for all four categories of collective

---

[1] We recognize that someone could be writing a letter to the editor or appearing at an EIS hearing as part of a collective effort to oppose or support the project; however, this is often difficult to determine after the fact from the text of the letter or hearing transcripts.

[2] There is a segment of the literature on the EIS process that argues that, because the EIS is designed to invite public participation and controversy into the formal decision-making process, it will thus depress informal controversy (Freudenburg 2004; Lesbirel and Shaw 2005). However, we believe and have shown in other work (McAdam et al. 2010) that consultation affords opposition groups another opportunity to protest. This is consistent with the general consensus among social movement scholars that open political systems are more prone to protest than are closed ones (Eisinger 1973; McAdam 1999).

TABLE 3.1. *Level of Mobilization Activities by Case*

Location	Letters	Speakers	Coordinated Appearances	Public Meetings	Lawsuit (*y/n*)	Protest Events
Ventura County, CA	246	138	5	3	n	8
Claiborne County, MS	2	9	1	2	y	1
Solano County, CA	192	120	1	3	n	4
Malibu, CA	92	138	6	3	n	5
Aiken County, SC	12	21	0	1	y	2
Mobile County, AL	47	38	0	1	n	1
New Castle County, DE	9	20	1	1	n	1
Cameron Parish, LA (Gulf Landing)	0	1	2	0	y	2
City of Long Beach	65	22	5	1	n	2
Riverside County	5	72	5	1	n	1
Essex County, MA	19	26	0	1	n	0
Gloucester County, NJ	6	13	0	1	n	0
Providence County, RI	15	17	0	1	n	0
Brazoria County, TX	5	7	1	0	n	0
Cassia County, ID	5	5	1	1	n	0
Cameron Parish, LA (Creole Trail)	0	1	0	0	n	0
Cameron Parish, LA (Sabine Pass)	0	1	0	0	n	0
San Patricio County, TX (Vista del Sol)	2	0	0	0	n	0
San Patricio County, TX (Cheniere)	0	0	0	0	n	0
Whatcom County, WA	2	4	0	0	n	0

action. Across all twenty cases, we recorded a total of just twenty-seven protests and three lawsuits. Applying a more holistic standard, a sustained social movement emerged in only two of our communities, Ventura County and Malibu, but even this is misleading because the two communities were acting in concert to oppose the same LNG project. So we are really talking about a single social movement here. Had we conformed to the methodological conventions of the field, we might well have picked this case for study, remaining blithely unaware of just how rare and atypical a case it is relative to our full sample of at-risk communities. Just how unrepresentative is it? Consider the extent to which this case – or really the combined cases of Ventura County and Malibu – dominates and distorts the descriptive statistics reported in Table 3.1. In Table 3.2, we compare the number of actions in each of our six categories that occurred in Malibu/Ventura County versus the remaining eighteen communities in the study.

The Malibu/Ventura County "movement" accounts for 47 percent of all the letters, 39 percent of the coordinated actions, and just short of half of all the protest events recorded for the study as a whole. How misleading it would be to try to understand community response to risky energy projects through the lens of the Malibu/Ventura County case.

Turning to the two criteria we use to define *opposition mobilization*, we find that ten communities experienced at least one – and often only one – protest and fifteen were judged to have seen some level of collective action. Because both criteria have to be satisfied for a community to be seen as having mobilized, we are left with a precise 50/50 split in our cases. That is, half of the communities "mobilized" and half did not. In the next section we explore this variation in considerable detail, using the technique of fs/QCA to try to predict mobilization and nonmobilization. But before we do

TABLE 3.2. *Malibu/Ventura County Mobilization Activity Compared to All Other Cases*

	Malibu/Ventura County	All Other Locales
Letters	338	386
EIS speakers	138	377
Coordinated actions	11	17
Public meetings	6	14
Protest events	13	14
Lawsuits	0	3

so, we want to close this section by moving beyond the statistics to briefly profile a case that strikes us as broadly modal in the overall level and forms of action it generated in response to the proposed project.

A TYPICAL RESPONSE: IDAHO WIND FARM

In the fall of 2002, the Boise-based company, Windland, Inc., announced plans to build a large "wind farm" on Cotterel Mountain in Cassia County, a large but sparsely populated rural county located in south central Idaho bordering Nevada and Utah. The "farm" was to consist of a cluster of up to 130 very large (210-foot tall) wind turbines that would generate electricity through wind power. The initial reaction to the announcement was almost universally positive. The paper of record, the Twin Falls *Times-News*, endorsed the project just five days after Windland held its first public meeting on the proposal. Elected officials lined up behind the project from the start. Even though it had no official role to play in the environmental review, the city council of the town closest to the proposed site – Albion, Idaho – issued a unanimous position paper in support of the project in March 2003. In supporting the plan, proponents stressed four main benefits:

- *Alternative energy*. Against the backdrop of increasing concern about our "dependence on foreign oil" (especially so soon after 9/11) and "global warming," supporters touted the project as an important step in the development of clean, alternative sources of energy.
- *Tax revenues*. Early in the review process, county officials announced that the project would mean an additional $639,000 in property taxes on an annual basis. In such a sparsely populated country like Cassia, this was widely regarded as a significant windfall and a major reason for backing the project.
- *Jobs*. Windland, Inc. estimated that the project would support eighty to one hundred jobs during the active construction phase of the project and a permanent staff of twenty to operate and service the turbines afterward.
- *Local pride/general development of the area*. Proponents also displayed a good bit of local "boosterism" in supporting the project, arguing that it would bring increased visibility and hopefully more development to the small and economically depressed communities in the area.

Despite all of these positives and the nearly universal support accorded the project at the outset, opposition to the proposal began to surface in the spring of 2004. The opposition was led by a relatively recent transplant from California who had moved to Idaho, in part, to escape the chaos and clutter of urban life and whose home looked out on Cotterel Mountain. Although our fieldwork led us to conclude that the great majority of area residents never wavered in their support for the project, the California expat was not entirely alone in his opposition. He formed the Committee Against Windmills in Albion and managed to get some seventy-five residents to attend a meeting sponsored by the committee in June 2004. Most of those who spoke at the meeting also voiced reservations about the project. The high point of the opposition "movement" came on July 7 of the same year when, under some pressure from the committee, the Albion City Council voted to rescind their endorsement of the project. It should be made clear that the vote did not mean that the council now opposed the project only that, in light of the recent opposition, certain council members felt uncomfortable as being on record as supporting the plan. The vote was meant to signal that the council was now neutral on the matter.

The only other significant opposition voice that surfaced during the EIS review process was a regional supervisor for the Idaho Department of Fish and Game who argued against the project, based on Cotterel Mountain's importance as a habitat and breeding ground for the sage grouse. The committee half-heartedly incorporated this argument into its discussion of the issue but generally steered clear of any strident environmental framing of the matter in deference to the conservatism of the area. The California expat admitted that he feared that being identified as an environmentalist and a Californian would be the kiss of death for the opposition! When it became known that the leader of the opposition was from California (aka, one of those contributing to the "Californication" of Idaho) and was seeking to block the plan largely to protect the view from his backyard, public opinion turned against the committee. After hitting its high-water mark in June and July 2004, nothing much was heard from the committee or organized opposition more generally for the duration of the decision-making process. The formal resolution of the matter came in August 2006. At that time the Bureau of Land Management – on whose land the wind farm was to be built – issued its report, granting a federal permit to Windland to proceed with the project. The Cassia County Planning and Zoning Commission followed suit in January 2007, issuing Windland a conditional-use permit for the wind farm.

So how much emergent opposition activity actually took place in this instance? Answer: very little. Besides the one public meeting called by the Committee Against Windmills in Albion, the case featured no other examples of coordinated collective action. No legal action was taken against the project. No protest events were staged in opposition to the plan. Even individual expressions of opposition were limited. Of the letters to the editor that addressed the issue, only five were from individuals opposed to the wind farm, and two of these were written by the afore-mentioned head of the committee. Of the mandated "comments" that were appended to the EIS report issued in August 2006, only five argued against the project.

We close this litany of relative inactivity to underscore the key point of this section. Remember that we chose to profile the Idaho case because its level of mobilization was seen as broadly modal for our sample as a whole. In general terms, the Idaho case falls in the middle of the continuum on this all-important "outcome condition." That it does this while exhibiting so little in the way of emergent collective action should tell us all we need to know about the exceedingly modest "baseline" of contention uncovered by this research.

EXPLAINING VARIATION IN MOBILIZATION/DEMOBILIZATION

Although true comparative case studies of at-risk populations are rare (though see Sherman 2011; Walsh et al. 1997), this does not mean that we lack for empirical work that bears on the issue of variation in emergent collective action. Even as we have criticized the tendency of scholars to ignore variation in mobilization in favor of single case studies of canonical movements, we are drawn to the social movement literature precisely because of its central concern with explaining emergence and collective action. Even if movement analysts have, from our perspective, consistently selected on what we see as the central dependent variable, they are still asking the same question that concerns us here. What factors encourage emergent mobilization? To begin to fashion a satisfactory answer to that question, it makes sense to turn first to the social movement literature.

Over the full history of social movement studies, various answers have been offered in response to the aforementioned question, but something of a theoretical consensus has been evident in the field for some fifteen to twenty years. Sherman (2011) refers to it as the "classic social movement agenda." That agenda reflects a kind of stylized or truncated version of the "political process model" proposed by McAdam in his 1982 book on the origins of the modern civil rights struggle. In that study, McAdam empha-sized the confluence of three causal factors: (1) changing political conditions that rendered the American state more vulnerable to challenge on the issue of race, (2) the presence of strong, segregated institutions that afforded civil rights activists the organizational infrastructure and resources needed to mobilize and sustain the movement, and (3) growing optimism within the black community that significant racial change was possible through concerted collective action. These factors are now better known by the shorthand designations of "political opportunities," "organizational capacity," and "cognitive liberation." Over time, however, a funny thing happened. The theory came to be identified almost exclusively with the structural components of the model – political opportunities and organizational capacity – while the key subjective/cultural dimension of the original formulation – cognitive liberation – was largely forgotten.

For reasons we will explain in the following text, we are critical of the truncated model and especially the absence of any serious attention to the emergent, constructed perceptions of threat or opportunity that furnish the necessary motivation for action. We will build these perceptions back into our causal "recipes" at a later stage of the analysis. Here, however, we stick with convention and include only the two structural variables in our provisional model. To these two canonical social movement variables, we add a third variable drawn from the literature on "facility siting." The literature on facility siting seeks to understand opposition to facility proposals and suggest better ways to site such facilities. The variable in question concerns the "risk" inherent in the proposed facility. Mirroring the "structural bias" in the "classic social movement agenda," however, the focus in the facility siting literature is on the objective risks associated with the project, rather than the subjective perceptions of risk (or reward) among those who would be affected by the project. We are as critical of this "objective" construction of risk as we are of the failure of most social movement scholars to assign central importance to collective meaning making and social construction during the all-important embryonic stage of emergent collective action. We will have much more to say about these crucial subjective processes later on in the chapter and throughout the book. For now, however, we will limit ourselves to the three variables discussed so far: political opportunities, organizational capacity, and objective risk. The question is simple: how far will these tried and true variables – or causal conditions in the language of fs/QCA – go in helping us understand variation in mobilization across our twenty communities?

Before we present the results of this initial analysis, we provide a brief overview of the fs/QCA technique. For each outcome, fs/QCA provides information about which causal conditions (or combinations of causal conditions) are necessary or sufficient to produce the outcome of interest. Necessary causal conditions are those that must be present but alone are not sufficient to produce the outcome of interest. Fuzzy-set scores of a necessary causal condition (X) are consistently greater than or equal to fuzzy-set scores of the outcome condition (Y) for most cases, or $X_i \geq Y_i$.[3]

Sufficient causal conditions (or combinations of causal conditions) are those that are sufficient but not necessary (because of multiple causal pathways) to produce the outcome of interest. Fuzzy-set scores of a sufficient causal condition (X) are consistently less than or equal to fuzzy-set scores of the outcome condition (Y) for most cases, or $X_i \leq Y_i$.

In addition, fs/QCA provides information about the consistency and coverage of individual causal recipes and the combination of causal recipes produced in an analysis. Consistency, the more important calculation, measures the degree to which one condition is a subset of the other. Consistency scores of greater than 0.8 for sufficient conditions (or causal combinations) and 0.9 for necessary conditions are commonly used as guidelines by scholars of fs/QCA in order to establish the relevant set-theoretic relationship between the causal and outcome conditions.[4] If X is determined to be a consistent subset of Y (or a sufficient condition of Y), then its coverage score reveals how important the combination of conditions represented in X is in accounting for Y.

As a brief reminder, our "classic" components of the social movement agenda and facility siting literature were measured as follows:

- *Risk* was scored as a combination of three measures of threat that a project poses (environment, safety, and property value).
- *Political opportunity* was scored as a combination of three measures of the openness of the decision-making process that determined the fate of the project. These three measures include (1) electoral vulnerability of decision makers (e.g., were the key decision makers elected officials?), (2) proximity to an upcoming election (e.g., were the key decision makers up for reelection during the course of the EIS process), and (3) jurisdiction (e.g., local, state, and federal, with *local*

---

[3] This relationship implies that the causal condition is a superset of the outcome. Similar to statistical analysis, it is important to remember that this mathematical relationship does not imply necessity without concrete causal evidence from our individual, real-world cases. Absent this concrete linkage, such a condition could simply be an attribute that is shared by instances of the outcome and may not be causal. The condition could be constitutive – essential in some definitional way to the outcome – or merely descriptive – something that the cases displaying the outcome just happen to share.

[4] C. Ragin. Personal communication to second author. February 2008.

defined as affording opposition the greatest "opportunity" for influence).

- *Organizational/civic capacity* was measured as a combination of the community's organizational capacity (operationalized as nonprofits per capita), political involvement (voter turnout), and mean education levels.

Using these components (drawn from the classic social movement agenda and facility siting explanations), our results prove underwhelming (see Figure 3.1). The recipe that results from this set of factors explains only three cases, which are, somewhat unsurprisingly, Malibu, Ventura County, and Mobile County. As described in the preceding text, two of these cases (Malibu and Ventura County) actually represent the response to a single project – Cabrillo Port. Moreover, the response to Cabrillo Port, which included a great deal of activity and several large protest events, most closely aligns with what social movement scholars would typically study. Mobile County's response to Compass Port was also highly contentious. Thus it is no surprise that the classic social movement agenda provides a consistent explanation of these cases. What is surprising is that

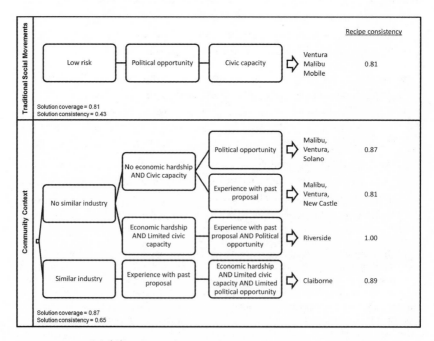

FIGURE 3.1. Mobilization recipes

our objective measure of project risk is low for Cabrillo Port and Compass Port, indicating that these communities mobilized despite facing a relatively low-risk proposal. We will have more to say about this later, but the bottom line is that the classic social movement agenda does little to help us understand the full range of responses among our twenty communities.

Our results mirror those obtained by Sherman (2011) in his study of opposition to radioactive waste-site proposals. Although Sherman used Poisson regression models rather than fs/QCA to analyze his data, in other respects his study matches ours closely. Two commonalities are worth noting. First, like us, Sherman picked locales for study rather than "movements" or mobilization attempts. Although we compare twenty communities slated for environmentally risky energy projects, Sherman (2011) looks at variation in "acts of public opposition to low-level radioactive waste site proposals across twenty-one counties" in the United States. Second, he too begins his analysis by assessing the predictive utility of the "classic social movement agenda."[5] Like us, he finds the agenda woefully inadequate to the task. There is, however, one variable in his model that does perform well. Using letters to the editor that were published during the siting process, Sherman created a proxy for what he calls the strength of the "injustice frame" in each of the twenty-one counties in his sample. Significantly, this variable bears a strong predictive relationship to "acts of public opposition" to the waste-site proposal. Sherman goes on to explain the significance he attributes to the variable. He writes,

This variable is the only independent variable in the classic agenda model that measures a phenomenon that is actively created during the siting process. Political opportunity structures and mobilizing structures are pre-existing conditions in the candidate counties. The significance of the framing variable, which measures a dynamic process during the episode of contention, should shift research and analysis away from pre-existing factors to dynamic aspects of the process that influence mobilization. McAdam explains the importance of framing relative to the other key factors of the social movement model this way. "While important, expanding political opportunities and indigenous organizations do not, in any simple sense, produce a social movement. In the absence of one other crucial process these two factors remains necessary, but insufficient, causes of insurgency .... Mediating between opportunity and action are people and the subjective meanings they attach to their situations." (1999: 48)

---

[5] We should make it clear that Sherman's variables are not identical to our causal conditions. Both of us employ proxies for political opportunity and organizational capacity, but while we then include a measure of objective risk in our recipes, Sherman adds a proxy for injustice frame in his model.

We heartily endorse the thrust of Sherman's argument. He has explained exactly why we are so critical of the "truncated" version of the political process model that is so widely accepted by social movement scholars. By essentially dropping the subjective, cultural component (e.g., cognitive liberation) out of the theory, analysts have not so much omitted one of three discrete variables as they have fundamentally abandoned the conjunctural logic that was central to the original perspective. The model was never meant to be additive – that is, political opportunities + organization + cognitive liberation = mobilization – but interactive. Shifting political conditions (e.g., political opportunities) were thought to be important partly because they encouraged "cognitive liberation" among an aggrieved population. "Mobilizing structures" were seen as important not only for the objective resources they commanded but also as sites within which the crucial processes of "social construction" and "collective attribution" could take place. In short, neither political opportunities nor mobilizing structures were ever thought to have direct effects on mobilization. Instead the presence of the two variables was thought to increase the likelihood of "cognitive liberation," which was seen as the true catalyst of emergent collective action. The key to increasing the predictive power of the social movement agenda, as Sherman notes, is to restore these crucial social/ cultural processes to their central place in our models of mobilization. To do that we must, in turn, be able to attend closely – in a methodological sense – to the specific geographic locales of interest. For it is within these locales that shared, emergent understandings shape the prospects for contentious collective action.

Bringing Communities Back in to the Study of Contention

Nearly a quarter of a century ago, McAdam, McCarthy, and Zald (1988: 729) wrote the following:

we [are] . . . convinced that the real action in social movements takes place at some level intermediate between the macro and micro. It is there in the existing associational groups or networks of the aggrieved community that the first groping steps toward collective action are taken . . . . Most of our research has missed this level of analysis. We have focused the lion's share of our research energies on the before and after of collective action. The "before" research has focused on the macro and micro factors that make movements and individual activism more likely. The "after" side of the research equation is composed of the . . . studies that focus on the outcomes of collective action. But we haven't devoted a lot of attention to the *ongoing accomplishment of collective action* . . . . [W]hat is needed is more systematic . . . fieldwork into the dynamics of collective action at the intermediate meso level. We remain convinced that this is *the* level at which most movement action occurs and of which we know the least. (emphasis in original)

The preceding characterization of the field seems nearly as relevant today as it was in 1988 when the original passage appeared. The truth of the matter is research on local contention is rare compared to the celebrated case studies of "national" movements that have largely defined the field (see, e.g., Adam 1987; Amenta 2006; Banaszak 1996; Beissinger 2001; Buechler 1990; Clemens 1997; Costain 1992; Ganz 2009; Gould 2009; Jenkins 1985; Kimeldorf 1999; Koopmans 1995; McAdam 1999 [1982]; McVeigh 2009; Meyer 1990; Morris 1984; Nagel 1996; Parsa 1989; Schwartz 1976; Smith 1991; Tarrow 1989; Taylor 1996; Tilly, Tilly, and Tilly 1975). For all their many virtues, these studies perpetuate a serious fiction, one we have underscored by placing the term *national* in quotes. In point of fact, virtually all "national" struggles are really aggregations of local movements. Mobilization may span a country, but it almost never takes place nationally. With the exception of the rare national strike or other coordinated action, mobilization is always embedded in and shaped by a local community context. That is why, from the beginning, we conceived of ours as a comparative study of at-risk *communities* rather than movements per se.

In focusing on communities, we certainly did not mean to ignore what McAdam, McCarthy, and Zald termed the "before" variables, that is, those features of the local context that were in place prior to the announcement of the proposed project. The goal of the first phase of our data collection was to create a static, comparative portrait of our twenty communities, believing that these "before" conditions would powerfully shape the popular perceptions of the project that emerged in the wake of the siting announcement. Besides "political opportunity" and "organizational or civic capacity," the key context variables for us include the following:

- *Economic hardship* was measured as a combination of unemployment rate and median income to reflect how much the community needed the injection of revenue and jobs associated with an infrastructure project.
- *Experience with prior land-use issue* was coded to reflect the existence, temporal proximity, and similarity of a previous experience opposing an infrastructure project or a similar land-use issue.
- *Similar industry* measured the presence of a similar industry in the community as a proxy for a community's familiarity and likely comfort with a technology similar to the one being proposed.

But again, the importance of these "context" variables stem, for us, from the influence they are expected to exert on the popular perceptions of the project that invariably emerge in the wake of any siting announcement. It should be clear that we have little direct evidence of these perceptions. Given the retrospective nature of our fieldwork, we simply were not in a position to directly observe the interpretive processes at work in our communities immediately following the siting announcement. The best we can do is infer the prevailing perceptions of threat from a combination of the "context" variables and letters to the editor written during the active EIS review period. That said, we feel confident in the predictions we have made regarding the link between these variables and the risks attributed to the projects.

All things equal, economic hardship is expected to decrease the likelihood that the citizenry would define the proposed project as a threat to the community. The same is true for similar industry. That is, we expect community familiarity with the type of facility or technology to be used in the project – or perhaps even the company that has proposed the project – to suppress "attributions of threat" by community members. Conversely, we see the chances of the community defining the project as threatening as greatly increased if it has previously mobilized against a similar land-use proposal. The question is, does adding these "context" variables to the classic social movement agenda improve our ability to explain variation in reactive mobilization across our twenty communities? The answer is a resounding yes.

As shown in Figure 3.1, the four recipes that result from combinations of these context-specific conditions explain six (of ten) cases that mobilized with high consistency. As expected, the most common route to opposition occurred in communities that did not have an industry similar to the one proposed, so they were unfamiliar with the technology. For these communities that did not have a similar industry, the recipe to opposition divides into two paths: (1) communities without economic hardship and with civic capacity – the most common path, explaining four cases – and (2) communities with hardship and without civic capacity. Oppositional recipes for communities in the first category also include either political opportunity (as in Solano County) or experience opposing a previous proposal (as in New Castle County) or both (as in Malibu and Ventura County). For those communities in the second category, political opportunity and experience are required for mobilization (as in Riverside County). These three recipes explain half of the cases that mobilized. An examination of the Cabrillo Port proposal in Malibu and Ventura County,

California, will highlight the key mechanisms at work in this set of cases. We will also contrast this case with the Northeast Gateway proposal in Essex County, Massachusetts, where all of these factors were also present but little mobilization occurred.

For mobilization to occur in a community with a similar industry, the recipe is somewhat surprising and includes economic hardship, limited civic capacity, and limited political opportunity. Of our hypothesized causal conditions, the only one that is present as expected is previous oppositional experience. A closer look at the Claiborne County case will illuminate other factors that may deserve attention in future analyses of community response. These same factors also help to explain the three cases that mobilized but are not explained by our recipes (Long Beach, Aiken, and Cameron Parish Gulf Landing).

A Closer Look at Some Cases

So much for the statistical analysis. But we didn't invest so much time and energy in systematically collecting four to six years of newspaper data and conducting interviews with the key players involved in each case just to give us an informed basis for assigning fuzzy-set values to our various causal conditions. Reflecting the hard lessons learned from the comparative study of water and pipeline projects described in Chapter 2, our commitment to conventional fieldwork here was motivated by nothing so much as a desire to amass a deep qualitative understanding of each of our cases. In bringing this section to a close, we want to draw on that knowledge to offer a holistic narrative account of two of the cases that are nicely captured by the fuzzy-set recipes discussed in the preceding text.

Malibu and Ventura County, California (Cabrillo Port)

In many ways, the Cabrillo Port proposal was up against the perfect storm of factors encouraging opposition to develop.[6] The proposal was located in an area that had little similar industry, had little economic hardship, possessed the civic capacity needed to mobilize, and had experience opposing the very same type of project only two years prior to Cabrillo Port. In addition, compared to other cases in our sample, the decision-making structures that would determine the fate of the proposal were more open to local influence. The risk posed by the facility was the one factor that was

[6] Adapted from Boudet and Ortolano (2010) and Boudet (2010).

lower for this facility than for others in our sample. The proposed facility
was to be located about fourteen miles offshore from Malibu and Ventura
County. Being so far from shore, Cabrillo Port seemed like the perfect way
to quell public-safety fears that had plagued earlier LNG terminal proposals in California. However, as we have argued in the preceding text, this
more objective measure of risk was trumped by risk perceptions determined not by the type of project but by the community context within
which the facility was proposed.

Prior to the Cabrillo Port proposal, Ventura County had already
rejected several attempts to site LNG facilities in its backyard. The first
LNG siting attempt in Oxnard came in 1974 when the Western LNG
Terminal Company[7] requested approval of three onshore LNG import
sites on the California Coast – at Oxnard, Point Conception, and the Los
Angeles Harbor. The Oxnard City Council was particularly active in
opposing the Western LNG Oxnard siting proposal and commissioned
an independent study of the impacts of the proposed terminal, which
concluded that a worst-case scenario could result in seventy thousand
causalities – a number that, not surprisingly, got the public to take notice
(Kunreuther and Linnerooth 1982). (Despite significant differences in the
facilities proposed in 1974 and 2003, this report resurfaced during the
Cabrillo Port debate, strengthening the hand of project opponents.) In
addition, the Sierra Club, which initially favored the Oxnard site in
1974, changed its position in 1977. The Sierra Club eventually joined the
opposition to Cabrillo Port as well. Finally, one of the first cases won by
the then newly created nonprofit law firm – the Environmental Defense
Center (EDC) – was representing the S. B. Indian Center in a suit against
the California Public Utilities Commission's decision to approve Western
LNG's Point Conception proposal (Environmental Defense Center 2002).
EDC was ultimately hired by the opposition to fight Cabrillo Port.[8]

A second attempt to site an LNG terminal in Oxnard came in 2002
when Oxy Energy Services, a subsidiary of Occidental Petroleum, attempted to purchase 265 acres of the Ormond Beach Wetlands to construct an
onshore LNG facility. Many of the same groups that originally opposed
the 1974 proposals and later opposed the Cabrillo Port facility – the

---

[7] Western LNG Terminal Company represented the interests of three utilities: Pacific
Lighting Corporation, Pacific Gas and Electric, and eventually, the El Paso Natural Gas
Company.

[8] After eight years of debate and despite being close to securing approvals, Western LNG
decided to defer construction of the proposed Point Conception facility to an unspecified
date because California no longer needed to import natural gas.

Oxnard City Council, Sierra Club, and EDC – were involved in a successful effort to push the Coastal Conservancy to purchase the property before Oxy could. In an article describing the events surrounding the Oxy proposal, Jane Tolmach, Oxnard's first female mayor, who had opposed the 1974 attempt and later voiced opposition to Cabrillo Port, was described as "ready to pull files down from the attic" (Gustaitis 2002) when Oxy came to town. California State Assembly members Fran Pavley and Hannah-Beth Jackson, Oxnard Mayor Manual Lopez, and Ventura County Supervisor John Flynn all spoke out against the Oxy proposal and later opposed the Cabrillo Port proposal. Thus prior to the Cabrillo Port proposal, a significant network of anti-LNG activists already existed in Ventura County. This network was readily available and easily mobilized against Cabrillo Port.

In addition to a history of successfully opposing LNG proposals, Ventura County was also the site of a more recent proposal by Crystal Energy to convert a defunct oil platform ten miles offshore of Ventura County into an LNG receiving terminal. This proposal was announced in March 2003, only seven months before Cabrillo Port, but was later stalled during the approval process. As a result of these experiences fighting previous LNG proposals, many community members in and around Oxnard were already well informed about LNG and easily mobilized to oppose Cabrillo Port. Many of the same groups and individuals who were involved in opposing previous LNG siting attempts in Oxnard quickly became involved in efforts to oppose Cabrillo Port.[9]

Unlike Ventura County, Malibu had not previously been involved in an LNG fight. At the same time, Malibu had waged significant battles over land-use and development issues throughout its history. Several interviewees from Malibu compared the fight over LNG to the battle for cityhood during the early 1990s.[10] There had been a few unsuccessful attempts to incorporate Malibu as a city in the 1960s and 1970s. However, pressures to become a city were reignited in the 1980s due to "fears of overdevelopment and resentment in Malibu of government far away in Los Angeles" (Chriss 2005a). Los Angeles County's plan to build a sewer system in Malibu, a move that most residents believed would lead to higher-density

---

[9] J. Flynn, interview with second author, May 2, 2007; C. Godwin and S. Godwin, interview with second author, July 17, 2007; T. Holdren, interview with second author, May 2, 2007; D. O'Leary, interview with second author, May 3, 2007; G. Roman and W. Terry, interview with second author, May 2, 2007.

[10] L. Griffin, interview with second author, July 18, 2007; H. Laetz, interview with second author, May 4, 2007; A. Stern, telephone interview with second author, March 20, 2007.

development, was the final straw. In 1990, the issue of cityhood went to a ballot and passed. Malibu was incorporated on March 28, 1991.

Since becoming a city, Malibu also fought a major battle with the California Coastal Commission on the formation of a Local Coastal Program (LCP), as required for coastal development under the Coastal Development Act. After six years of debate, which left Malibu's coastal development at a virtual standstill, the state legislature passed AB988 in 2000, essentially preempting local control in Malibu by requiring that the California Coastal Commission draft Malibu's LCP. Lawsuits ensued. It was not until August 2005 that the issue was somewhat resolved, with the California Coastal Commission agreeing to develop an LCP in concert with Malibu (Chriss 2005b). Interestingly, many of the same people who were on opposite sides of the LCP debate joined together to oppose the Cabrillo Port facility. Thus despite a lack of experience with LNG proposals, Malibu has a long and storied history of fighting development proposals, and because of its wealth and Hollywood celebrities, it has the resources and connections to do so.

BHP Billiton's Cabrillo Port proposal first went public when the Australian company submitted applications to the Federal Maritime Administration for a deepwater port license. The application triggered a federal EIS, orchestrated by the U.S. Coast Guard, and a state Environmental Impact Report (EIR), under the direction of the State Lands Commission. Its offshore location also meant that then California Governor Arnold Schwarzenegger would have veto authority over the proposal. The involvement of state officials in the approval process created a key political opportunity for opponents.

There was no major press conference to announce BHP's Cabrillo Port proposal. Instead, residents learned of the plan through newspapers. Despite this relatively low-key announcement, many politically active Ventura County residents became aware of the facility because they were monitoring energy industry activities on the West Coast as a result of their previous fights against LNG. These individuals and organizations quickly mobilized against the facility, contacting local elected officials, picketing outside BHP's headquarters in Oxnard, and testifying at public meetings. However, they found it difficult to spread their opposition beyond Oxnard.[11] In Malibu, by contrast, the proposal initially stayed largely "under the radar."[12] Although the Malibu City Council quickly passed a

[11] Godwin and Godwin interview, July 17, 2007; Roman and Terry interview, May 2, 2007.
[12] Stern interview, March 20, 2007.

resolution against the proposal, few residents were actively engaged prior to the release of the Revised Draft EIR.

With the release of the Draft EIS/EIR in November 2004 and an article in the *Los Angeles Times* soon thereafter, an established environmental group – the Sierra Club Great Coastal Places Campaign – became aware of the project and allied itself with the initial Oxnard opponents.[13] With the help of a paid community organizer from the campaign, Oxnard activists began holding regular meetings to plan opposition events. They held their first large-scale protest – "Hands along the Pipeline" – in May 2005. This event marked the end of the initial opposition group's small weekly protests outside BHP's offices in Oxnard and the beginning of a larger mobilization effort in the city. With the Sierra Club now heavily involved in shaping opposition strategy, opponents started to implement many of the same tactics as the opponents in Vallejo – "tabling" at the farmers market and festivals, canvassing neighborhoods with flyers and fact sheets, and organizing larger letter-writing and petition campaigns.

In addition, around this same time, Susan Jordan, a well-known coastal advocate and director of Santa Barbara's California Coastal Protection Network, joined the opposition. She had previously been focusing her efforts on enacting state legislation that would require a ranking of all LNG proposals prior to approval. However, when Cabrillo Port emerged as the front-runner in the race to site a facility, she decided to become involved in opposing the proposal.[14] Moreover, Ozzie Silna, a wealthy philanthropist in Malibu who had worked with Jordan before on Malibu's Local Coastal Plan, began backing her efforts to block Cabrillo Port. With money from Silna, Jordan paid for a review of the Draft EIS/EIR by the EDC, a nonprofit law firm in Southern California that Jordan had also worked with previously.[15] The EDC brought to the opposition the technical and legal know-how to review large, complicated regulatory documents, and Jordan brought the political skills and connections required to lobby state officials and the governor effectively. A strong coalition against Cabrillo Port, involving the leadership of the California Coastal Protection Network, EDC, Sierra Club, and Oxnard environmental and social justice groups, was beginning to take shape. However, opponents still struggled to mobilize the general public, particularly in Malibu.[16]

[13] O. Bailey, telephone interview with second author, May 3, 2007.
[14] S. Jordan, telephone interview with second author, November 20, 2007.
[15] Jordan interview, November 20, 2007; Laetz interview, May 4, 2007.
[16] Bailey interview, May 3, 2007; Griffin interview, July 18, 2007; M. Morales, telephone interview with second author, May 30, 2007; Stern interview, March 20, 2007.

Things began to change in late June 2005. A series of events and articles began to raise questions about the adequacy of the review process and the neutrality of decision makers. First, Governor Schwarzenegger commented after a speech that LNG was necessary for the state to meet its future energy demands and that his "personal preference" for a site was Oxnard (Herdt 2005). In addition, the *Malibu Times* published an article in June 2005, indicating that some fifty of the letters filed electronically in support of Cabrillo Port during the Draft EIS/EIR process had been falsified. Moreover, in a July 2005 article, the *Malibu Times* revealed that the U.S. Environmental Protection Agency (EPA), under pressure from the Bush administration (at the request of BHP), had reversed its earlier decision concerning the applicability of Ventura County's New Source Review requirements to the Cabrillo Port proposal. Originally, the EPA had determined that these requirements would apply to Cabrillo Port, thus necessitating that BHP purchase emission-reduction offsets for the facility's air-pollution emissions, a difficult task considering a lack of offsets for sale in Ventura County. However, in June 2005, the EPA had quietly issued a letter exempting the project from these requirements by locating it in a different jurisdiction. Opponents, particularly the EDC, were quick to pick up on this change and submitted a series of Freedom of Information Act requests about the EPA's determination. The results of the EDC's investigations eventually led the Ventura County Air Pollution Control Board to challenge the EPA reversal in late 2006 and a led to a federal investigation by Congress in early 2007.[17]

Another blow against Cabrillo Port proponents came in the form of an August 2005 article in the *San Francisco Chronicle* that disclosed close ties between Governor Schwarzenegger and the law firm that BHP had hired to lobby on its behalf. Then a September 2005 *Ventura County Star* article revealed that BHP had spent $1.3 million in California on lobbying during the first six months of 2005. This figure placed the company third in the state in terms of money spent on lobbying during that period. These events during the summer and fall of 2005 galvanized the opposition in Oxnard and, more importantly, Malibu. In July 2005, the Malibu City Council unanimously approved a resolution requesting that the California attorney general investigate the false letters supporting the project in the Draft EIS/EIR. The city of Malibu also determined that its shoreline was closer to the facility than Ventura County and Oxnard. As a result, project opponents

---

[17] Jordan interview, November 20, 2007; State official, telephone interview with second author, April 25, 2007.

began actively to seek out support in Malibu and beyond. In September 2005, opponents sent nearly twenty-five thousand mailers to selected voters along the coast between Santa Barbara and Santa Monica. Also, Jordan's California Coastal Protection Network and the Sierra Club began to solicit involvement in fighting Cabrillo Port from movie and TV stars in Malibu who had historically been sympathetic to environmental causes.[18]

Pierce Brosnan (of James Bond fame) and his wife Keely Shaye Smith were among those contacted. As a result of Malibu's increased interest, the hearing on the Revised Draft EIR in Malibu became a raucous affair. BHP representatives and supporters were booed loudly during their testimony, and state officials contemplated shutting down the meeting prematurely. Although the Brosnans could not attend personally, they sent their personal assistant, Jolene Dodson. Impressed by the public outcry against the facility expressed at the Malibu hearing, Dodson introduced herself to Jordan. This liaison led to a subsequent decision by the Brosnans to join the fight, thereby giving the opposition highly visible allies.[19]

After several other movie stars joined the opposition, events became even bigger, incorporating Malibu and Oxnard opponents, and garnered increased media attention. Opponents planned an outdoor screening of *An Inconvenient Truth* in Malibu that drew six hundred people, a "paddle out" protest[20] that drew more than two thousand participants, and a rally at the final EIS/EIR hearing in Oxnard that also drew more than two thousand people. These events were covered in the entertainment and news media, locally and internationally. At the same time, Jordan and the EDC continued to review regulatory documents and present scientific and technical arguments against the facility to state officials. The project's downfall was sealed during the November 2006 elections. The Democrats' sweep of the U.S. Congress placed long-serving California Democrats in positions of power. This allowed opponents to bring their concerns to federal officials, who launched the previously mentioned congressional investigation of the EPA reversal. In addition, two of the three members of the California State Lands Commission, charged with reviewing the EIR, were newly elected in November 2006. These same two individuals

---

[18] Jordan interview, November 20, 2007; J. Dodson, interview with second author, July 18, 2007.

[19] Dodson interview, July 18, 2007; Griffin interview, July 18, 2007; Jordan interview, November 20, 2007.

[20] A paddle out is typically used by surfers to honor the life of a fallen surfer and involves a number of surfers paddling out in his or her memory. In this case, the paddle out symbolized the mourning of the death of the ocean if Cabrillo Port were to be built.

eventually voted to deny certification of the EIR in April 2007, marking the beginning of the end of the Cabrillo Port proposal. Several days later, the California Coastal Commission voted against the project, and finally, Governor Schwarzenegger vetoed the project in May 2007.

As this case narrative shows, the Cabrillo Port proposal confronted many factors that are associated with mobilization from the perspective of the classical social movement agenda (notably civic capacity and political opportunity) and our revised version of the theory that incorporates aspects of the community context (little similar industry, little economic hardship, and experience opposing the same type of project). It is these community context factors that provide the spark for mobilization, not the more traditional aspects of threat (or risk) that are often included in research by scholars attempting to understand the Not in My Backyard (NIMBY) response.

Interestingly, there is another case that fits into the same causal recipes as Malibu and Ventura County in our analyses but failed to generate a great deal of opposition: Essex County, Massachusetts. Similar to Cabrillo Port, the Northeast Gateway proposal in Essex County was located far offshore from Gloucester in an effort to overcome the public-safety fears that had plagued onshore LNG proposals in Massachusetts. Unlike the California case, however, the local opposition in Gloucester, which was largely comprised of fishermen who did not want to lose valuable fishing areas to the facility, was unable to attract broader regional support. There were several reasons for this. First, regional environmental groups in the Northeast, such as the Conservation Law Foundation, held the view that natural gas importation for the Northeast was necessary to replace more polluting coal-fired electricity generation. The Northeast is much more dependent on coal as an energy source than California. Thus local opponents in Gloucester lacked regional allies like those that became involved in the Cabrillo Port case. Moreover, the most contested LNG proposal in the Northeast, Hess's Weaver's Cove Energy Project was located nearby and onshore in Fall River, Massachusetts. Opponents in Fall River – the Coalition for Responsible Siting of LNG Facilities – argued that LNG should be sited away from heavily populated areas and ideally offshore. The Fall River opponents' chosen framing of their opposition as "responsible siting" meant that allying with Gloucester opponents against an offshore facility was impossible. In California, communities facing different LNG proposals were able to coalesce into a regional alliance against the importation of LNG more generally, as opposed to more community-specific concerns about safety impacts. Thus we find that a combination of factors, external to the movement actors, made expanding opposition to Northeast Gateway beyond Gloucester difficult. These factors include energy supply concerns, the behavior of external regional organizations, and choices made by competing communities.

Claiborne County, Mississippi (Grand Gulf)

The proposal in Claiborne County, Mississippi, was for a second nuclear power plant at the existing Grand Gulf facility. Claiborne County is a predominantly African American county (84% according to the 2000 Census). As one of the poorest and least educated communities in our sample, Claiborne County did not score high on our measure of civic capacity. However, the community has a long history of civil rights activism, and the local chapter of the National Association for the Advancement of Colored People (NAACP) has been a major organizing force in the county. African Americans in Port Gibson, with help from the NAACP, staged a boycott of white merchants from 1966 to 1972 in order to gain equal rights within Port Gibson and Claiborne County (Crosby 2005). The local chapter of the NAACP also became involved in the Grand Gulf siting (discussed in the following text). For this reason, our measure of civic capacity, which focused on education levels, voter turnout, and environmental/health organizations, did not necessarily capture the type of activist potential present in Claiborne County.

In addition, the long tradition of activism against nuclear proposals in the United States – dating back to before the Three Mile Island disaster in 1979 – has left us with a legacy of established organizations committed to opposing any new nuclear facility regardless of where they are proposed in the country. Not surprisingly, then, external antinuclear groups became involved almost immediately in both communities in our sample in which nuclear facilities were proposed (Claiborne and Aiken). These national groups monitor the *Federal Register* for proposals and were on high alert prior to the Grand Gulf proposal given then Secretary of Energy Spencer Abraham's February 2002 announcement of the Nuclear Power 2010 Program. In Claiborne, this outside support took the form of the Nuclear Information and Resource Service (NIRS) Reactor Watchdog Group, Public Citizen, and Association of Communities Organizations for

---

[21] The Clamshell Alliance is famous for its protracted (1976–90) campaign against a New Hampshire nuclear facility that kicked off the antinuclear movement in the United States. The Alliance became well-known for their use of nonviolent civil disobedience. In April 1977, more than 2,000 Clamshell protestors occupied the Seabrook Nuclear Power Plant

Reform Now. NIRS, through Paul Gunter, a native of Mississippi and one of the founders of the Clamshell Alliance,[21] was particularly involved. NIRS and Public Citizen conducted site visits in the community, organized at least two public meetings with the local chapter of the NAACP, collected affidavits[22] from local first responders about the inadequacy of current emergency plans in the County, wrote up accounts of the proposal in press releases and newsletters, testified at the scoping and Draft EIS public meetings, met privately with state legislators, and organized rallies around EIS meetings. At the rally on the steps of the Capitol Building in Jackson before the meeting on the Draft EIS, opponents brought an ice sculpture of a nuclear reactor to symbolize the likely meltdown of the proposed facility. In addition to these political activities, opponents also worked to gain standing (which required participation from local residents) in the Nuclear Regulatory Commission's adjudicatory proceedings. They also secured legal representation from Diane Curran, who had worked with the same external groups to stop a uranium enrichment plant in Homer, Louisiana – one of the first successful environmental justice claims brought to the Nuclear Regulatory Commission. Many of these activities, which to a large extent were spurred by external groups, are what we measured in our count of mobilization.

During our interview, Gunter talked about how he believes in using the "full gamut" of tactics to oppose a facility: "You can't just rely on the Nuclear Regulatory Commission's licensing process. You have to enter the court of public opinion and use the media to get your message across." He sees all of these tactics as "legs on a chair" or "rungs on a ladder" for opponents to be successful. He stressed that this "integrated approach" is necessary to combat the "very powerful and influential" nuclear power industry because "the companies are not relying on the licensing process alone either. They're trying to integrate themselves into the community. They give money to charitable organizations. They back political candidates through political action committees. They lobby on Capitol Hill and hire prestigious law firms."[23] Knowing this information about the involvement of external groups in nuclear projects makes it less surprising that

construction site. A total of 1,414 of these activists were arrested and held in jails and National Guard armories for up to two weeks after refusing bail. This tactic, which garnered a great deal of media attention, was later copied by opposition groups in other areas, like Diablo Canyon and San Luis Obispo in California.

22 Affidavits were collected as input into eventual contentions submitted as part of the Nuclear Regulatory Commission's adjudicatory process. Collecting affidavits is a typical activity of NIRS in communities facing licensing proceedings.

23 P. Gunter, telephone interview with second author, March 3, 2009.

Claiborne County received a high score in terms of mobilization and suggests that future studies should include this factor as a causal component. In Claiborne County, these external groups overcame the impact of limited civic capacity and political opportunity structures on opposition mobilization by providing guidance and support to local communities faced with complicated regulatory structures overseen by faraway federal, appointed officials. All four cases that were not explained by our recipes received significant help from outside groups in organizing their contentious events.

Although Claiborne County scored in the set of mobilized communities (mainly due to the activities of external groups), many interviewees felt local opposition was muted compared to previous disputes. One local resident commented, "If the community had had issues with this proposal, you would have heard about it." As an example, another land-use dispute was occurring around the same time as the Grand Gulf proposal. Since 1987, the Mississippi Department of Transportation (MDOT) has mandated expansion of Highway 61 to four lanes in all locations. The segment of Highway 61 that runs through Port Gibson (the closest community in the county to the Grand Gulf facility) is called Church Street and is home to many beautiful historic churches and homes. It is also the only section of the entire highway for which a plan for expansion to four lanes has not been determined. According to interviewees, the debate over the MDOT project has been much more heated than anything related to the nuclear proposal. In addition, several interviewees pointed out that the community of Claiborne County definitely knows how to make its voice heard. During the civil rights struggle in the county, many residents actively protested in a situation that presented significantly more potential for personal harm as a result of their actions than mobilization against Grand Gulf.[24]

Although external groups were interested in preventing the construction of an additional nuclear plant, most local opponents were supportive of an additional reactor but were motivated to make sure that the local community received its fair share of the tax revenue that such a project would generate. Economic issues were a major concern in Claiborne County. Several people commented during interviews and at meetings on the EIS that they would hate to see what Claiborne County would look like without the tax revenue provided by the original Grand Gulf reactor. The editor of the local paper wrote in a column, "I tremble to think what sort of

[24] A. C. Garner, interview with second author, April 7, 2009; J. Johnston, interview with second author, April 7, 2009; K. Ross, interview with second author, April 7, 2009.

tax base Claiborne County would have today if Grand Gulf had not been built" (Crisler 2009). Residents specifically compared Claiborne to its more impoverished neighbor to the south, Jefferson County, where, according to the 2000 census, per capita income was $9,709 and 36 percent of the population lived below the poverty line. The concern about the distribution of tax revenues became the "lynchpin" that tied together external and local opponents[25] and was very much rooted in the community's past experience. However, this concern is only somewhat captured by our hypothesized causal condition of "economic hardship."

Originally, all of the property tax revenue from the first Grand Gulf nuclear power plant (at the time, $16 million) went to Claiborne County. This revenue represented an enormous sum for the county, which had previously operated on a budget of less than $1 million. However, in 1986, just one year after the first Grand Gulf plant began commercial operations, the Mississippi State Legislature passed a bill that divided this revenue among all the counties receiving power from the plant. Claiborne County sued the Tax Commission in response to the amendment (*Burrell v. Mississippi State Tax Commission*, 1988) but lost in a ruling by the Mississippi Supreme Court. This change to the tax code represents a lingering sore spot for the community. Several theories were advanced by our interviewees to explain the change in the tax code. One, offered mainly by white residents, is that newly elected black officials in the county mismanaged the resultant tax revenue[26] from the first plant, forcing the state legislature's hand. The county tax assessor at the time, Evan Doss, one of the first African Americans elected to this position in the state, did go to federal prison for embezzlement. A second theory, offered mainly by the black residents, considers the act by the state legislature as just another in a long line of discriminatory policies aimed at primarily African American counties in the state. As one resident put it, "the reason behind the change in the tax code in 1986 was because Claiborne County is a predominantly African American county, and I will continue to believe this until it's proved otherwise."[27]

Given Claiborne County's long history of civil rights activism, it is not surprising that these questions about environmental justice, particularly with

---

[25] Gunter interview, March 3, 2009.

[26] One interviewee suggested that Mississippi Power and Light (the original owner of Grand Gulf) had sided with Claiborne County on the tax issue, but when the county decided to raise the millage rate, the company withdrew its support. Without this powerful ally, the county was left at the mercy of the state legislature's decision regarding the tax code.

[27] Garner interview, April 7, 2009.

respect to the tax code, became a major issue. At the meeting on the Draft EIS, Evan Doss argued that the distribution of tax payments from the first facility

> is racist and, in fact, discriminates against the predominantly black Claiborne County .... Given the severity of the State of Mississippi's misconduct, it would be unreasonable in the extreme for the United States Nuclear Regulatory Commission to overlook the obvious and neglect to take appropriate measures to prevent further actual discrimination against the predominantly black Claiborne County in connection with the second Grand Gulf nuclear power plant. (Nuclear Regulatory Commission 2005)

A final theory for the redistribution came from Gunter. He believed, although he found no paperwork to prove it, that the impetus for the change in the tax structure was the anticipated rate shock to consumers that would have occurred because of cost overruns and delayed construction at the original Grand Gulf facility. Thus Gunter argued that, to lessen the rate shock to consumers in Western Mississippi over the cost of power from Grand Gulf, the state legislature used taxes that should have gone to the black community to subsidize some forty-five other communities that would receive energy from the facility.

The tax distribution issue in Claiborne County raises a larger question that relates to two other projects that were not explained by our recipes – do different causal conditions and mechanisms explain opposition aimed at modifying project proposals versus those that seek complete rejection of a facility? In the case of the Cameron Parish's Gulf Landing LNG projects in the Gulf of Mexico, the main goal of opponents was to "close the loop" and disallow the use of seawater in warming the LNG, not to do away with the projects altogether. Similarly, in the case of the Long Beach LNG Import Terminal, much of the opposition was aimed at moving the proposal offshore to limit potential safety impacts. This difference in goals among opponents in different cases and between local and external groups deserves further exploration in future research.

LEARNING FROM NEGATIVE CASES: TOWARD AN UNDERSTANDING OF NONMOBILIZATION

As we have noted several times, with rare exceptions, social movement scholars have sought to understand the onset of contention by studying

---

[28] We are heartened by the fact that in just the last few years a number of scholars have turned their attention to instances of nonmobilization. These include works by Shannon Bell (2010), Daniel Sherman (2011a, 2011b), and the exceptional 2009 book, *Flammable*, by Auyero and Swistun.

those exceedingly rare instances in which insurgent groups managed not only to generate collective action but also to sustain it over some extended period of time.[28] We have already commented on how this methodological convention almost certainly exaggerates the frequency and impact of social movements. There may, however, be a far more serious cost to this approach than simple exaggeration. Perhaps our basic understanding of the dynamics of contention is distorted by only studying such rare and clearly nonrepresentative cases. At the very least, it seems like an odd and risky bet to imagine that we can understand a phenomenon only be interrogating wildly successful instances of it. Would anyone ever propose studying religiosity only by investigating the lives of saints? Would it make sense to try to understand marital dynamics by interviewing only those couples that made it to their fiftieth anniversary? The received wisdom among methodologists is that one is apt to learn as much from "negative cases" as from positive instances of the same phenomenon. Unfortunately, the methodological conventions in the field have more or less precluded this possibility. Not in our case. As we reported in the preceding text, exactly half of our communities did not, by our criteria, mobilize in response to the proposed project. We want to exploit this fact by using the technique of fs/QCA to predict nonmobilization.

As with our analysis of mobilization, the community context conditions (similar industry, experience with past proposal, and economic hardship) do better at explaining nonmobilization than components of the classical social movement agenda (risk, civic capacity, and political opportunity), although the difference is not as overwhelming as with mobilization. Using the classical social movement factors, we find two recipes for nonmobilization as shown in Figure 3.2: (1) the combination of a low-risk project proposed for a community with limited civic capacity and (2) the combination of a high-risk project proposal with little political opportunity. The scores for consistency (0.80) and coverage (0.75) of these two recipes are reasonable. However, the consistency scores that result from combinations of context-specific conditions are much better (0.98) without much of a drop in coverage (0.73). As shown in Figure 3.2, the three recipes that result from combinations of context-specific conditions explain nine (of ten) cases that did not mobilize. The only case of nonmobilization that these three recipes fail to explain is Essex County, Massachusetts, which did not mobilize with the level of intensity expected for reasons explained in the preceding text.

As expected, the most common route to nonmobilization occurred in communities that had an industry similar to the one proposed, so they were

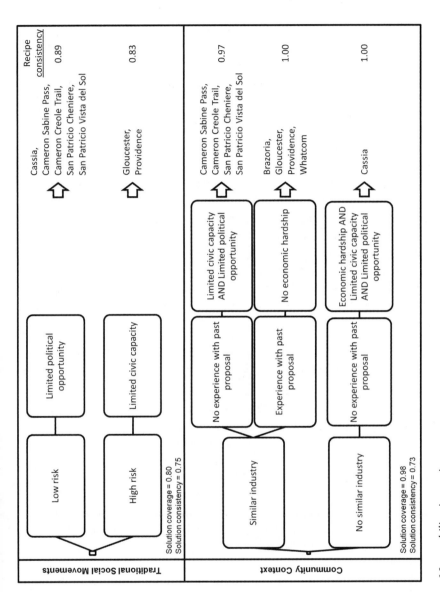

FIGURE 3.2. Nonmobilization recipes

familiar with the technology and, we suspect, inured of whatever risks it posed. For those communities that had experience with a similar industry, the recipe to nonmobilization divides into two paths: (1) communities with no experience opposing a past proposal, limited civic capacity, and limited political opportunity – an expected path, explaining four cases – and (2) communities with experience opposing a past proposal and no economic hardship – a somewhat unexpected path (apart from the presence of similar industry) that also explains four cases. A narrative analysis of two onshore LNG proposals in the Gulf Coast (Cameron Sabine Pass and Brazoria) will highlight the key mechanisms at work in this latter set of cases. We will also contrast the cases of Cameron Sabine Pass with Aiken County, where all of these factors were also present but mobilization nonetheless occurred.

For no mobilization to occur in a community without a similar industry, the recipe requires a number of other factors that make mobilization difficult, including no experience opposing a previous proposal, economic hardship, limited civic capacity and limited political opportunity. The previously described wind farm proposal in Cassia County is the one example that fits into this recipe.

As before, we want to illustrate and enhance our understanding of the recipes by digging a bit deeper into two of our "negative cases" – Cameron Sabine Pass and Brazoria.

Cameron Parish, Louisiana (Sabine Pass)

Louisiana has a long history of accepting and encouraging oil and gas development on and near its shores (Gramling and Freudenburg 1996). According to estimates by the Governor of Louisiana, the state provides 30 percent of the nation's petrochemical needs. Residents are also acutely aware of the positive economic impact these facilities bring to the area. Several interviewees explained how the oil and gas industry is a major contributor to the Louisiana economy.[29] Many described the industry as part of the "culture" of Louisiana, along with hunting and fishing. For example, one interviewee grew up with a gas plant in his backyard. At the time of the proposal, Cameron Parish was already home to nine pipeline processors, and nearby Calcasieu Parish housed twenty-three petrochemical facilities, which provided six thousand jobs. Despite this level of industrial development, no

---

[29] C. Griffith, interview with second author, January 12, 2009; G. Swift, interview with second author, January 12, 2009; S. Trahan, interview with second author, January 14, 2009.

major incidents associated with the oil and gas industry stuck out in people's memories. The community saw LNG as a natural extension of the oil and gas industry and a way to save Louisiana's and the Calcasieu Parish's dwindling economic prospects. For example, at the Federal Energy Regulatory Commission (FERC) hearing about the Draft EIS for the Sabine Pass project, Bobby Conner, the local tax assessor, asserted that

> Cheniere Energy ... would be a great help to our tax base, which is dwindling ... because ... everybody's going deeper and deeper offshore to drill for oil and gas these days. And the LNG facility, it would be real welcome .... Since 1999, Louisiana has lost 4,000 jobs in the petrochemical industry and is on track to lose another 18,000 in 2004–2005. Nowhere have they been stinged [*sic*] more sharply than the ammonia fertilizer industry where natural gas makes up 90 percent of the process. Four years ago there were nine aluminum plants in Louisiana. Today we have three. This has made the LNG project a top priority in our state and in our nation. (Federal Energy Regulatory Commission, 2004a: 22)

Not only were residents of Cameron Parish comfortable with oil and gas development generally at the time of the Sabine Pass proposal, but they also knew about LNG terminals specifically. Since 1981, nearby Lake Charles, Louisiana, had served as home to one of only four existing LNG facilities in the United States. Although interviewees mentioned that there had been some concerns associated with the original proposal for the Trunkline LNG facility, it had operated without significant incident since its inception. As a result, residents had grown accustomed to its operations and did not fear LNG. Kristi Darby, of Louisiana State University, commented in a *Mobile Press-Register* article that, "People in Louisiana are used to having the petroleum industry around. The oil industry has been in Louisiana for decades. The communities are not afraid" of LNG (James 2005). The Southwest Louisiana Partnership for Economic Development, then a newly created public entity whose board included elected representatives from each of the five parishes it served (including Cameron), was actively seeking industrial development before Sabine Pass. They were advertising that they had land available for industrial development in various economic development journals, including *Global Corporate Xpansion, Go Gulf, Trade and Industry Development, Expansion Solutions*, and *Site Selection*. Because residents had been so supportive of past industrial development, they had no experience opposing previous proposals. In addition, they had little civic capacity to do so – scoring low on education levels, organizational capacity, and voter turnout.

Site consultants, who had been hired by Cheniere, originally approached elected leaders and development officials in the community during the late

1990s about the possibility of developing an industrial facility in Cameron Parish. The use of site consultants is typical in this area. Cheniere had been looking across the river in Port Arthur, Texas (where ExxonMobil eventually sited their Golden Pass LNG facility), but the community and tax structure in Texas were not as hospitable as Louisiana. At the same time as the Sabine Pass proposal, other LNG proposals were on the table, including an expansion of the Trunkline LNG facility, Hackberry LNG (later Sempra's Cameron LNG), Shell's Gulf Landing, and ExxonMobil's Golden Pass (located in Sabine Pass, TX). It is worth noting that, just as communities in California and Massachusetts were fighting to delay siting of LNG proposals, communities in Louisiana and Texas were fighting to attract such proposals. This contrast in response to similar proposals provides important evidence of the importance of community context in shaping community mobilization around siting proposals.

James Ducote, a former employee of Cameron Parish's and the state's economic development departments (who now works for Cheniere Energy), was particularly instrumental in convincing Cheniere's Founder and CEO, Charif Souki, to locate in Cameron Parish. By all accounts, the original deal for the facility was struck in the back of a pickup truck. Ducote took Souki to visit what would eventually become the site of the Sabine Pass facility with the landowner, Butch Crain. The three shook hands on the deal during this visit. This handshake deal among company, state officials, and private landowner before any public consultation could occur contrasts sharply with what we might think of as a politically desirable process of decision making.

Given Cameron Parish's historically cozy relationship with the industry, it should come as no surprise that the community was very supportive when Cheniere filed its official application for the facility with FERC in December 2003. A local police juror (the equivalent of a county commissioner) was quoted in the local newspaper as saying to Cheniere representatives at a public meeting immediately following the proposal that, "All of Cameron Parish is behind your project. When you called, I thought we had got a call from heaven" (Wise 2003). The filing began the approval process, which included two hearings in the parish. Both were well-attended, drawing one to two hundred residents. All comments were supportive of the facility, except for one lone naysayer, Jerry Norris, a local fishing guide from the Texas side of the river, who was concerned about the placement of spoils from dredging associated with the construction of the facility. However, he was eventually persuaded to support the facility. In addition, a Cheniere representative estimated that the company held about two community meetings per week for two years, or almost two hundred meetings, over the course of the siting and approval process. The facility was approved less than one year after Cheniere submitted its application in December 2004.

The three other proposals that fall into this causal recipe (Cameron Creole Trail, San Patricio Cheniere, and San Patricio Vista del Sol) follow similar storylines. In many ways, the proposals that fall into this causal recipe (similar industry, no experience opposing a previous proposal, limited civic capacity and limited political opportunity) were not merely accepted but actively sought out as a way to spur local economic development. What little mobilization occurred in these cases was in support of the proposed facilities.

As mentioned in the preceding text, the Savannah River Site's (SRS's) Mixed Oxide (MOX) fuel fabrication facility in Aiken County, South Carolina, is an exception. This case falls into the same causal recipe (similar industry, no experience opposing a previous proposal, limited civic capacity, and limited political opportunity) but mobilized. However, similar to the other nuclear-related proposal in our set, all of this opposition came from outside the local community. The story of the local response to the proposal is remarkably similar to the cases in Cameron Parish and San Patricio County.

The SRS is located in the southwestern portion of the state of South Carolina, adjacent to the Savannah River, along the state border with Georgia, approximately twenty kilometers (12 miles) southeast of Aiken, South Carolina, and twenty-four kilometers (15 miles) east of Augusta, Georgia. The U.S. government owns the SRS, which was set aside in 1950 for the production of nuclear materials for national defense. National priorities have shifted since the end of the Cold War in 1991, and the site's priorities are now focused on waste management, environmental restoration, technology development and transfer, and economic development. The SRS covers about 803 kilometeres^2 (310 miles) in a circular tract of land within Aiken, Barnwell, and Allendale counties in South Carolina.

As in Cameron Parish and San Patricio County with the oil and gas industry, Aiken County residents link nuclear development at the SRS with economic development and were highly supportive of the SRS proposal. During one interview, an appointed official from the state government said that Aiken residents are pronuclear because they associate nuclear facilities with economic opportunities. He said that if a poll question were worded to include the word *nuclear*, the local community would reveal strong support, whereas if it merely described the work to be done, for example, storage of plutonium, local opinion might be more ambivalent. As a result of this association between nuclear and economic development, elected officials at the local (Aiken County) and state (South Carolina and Georgia) levels are consistently supportive of new nuclear facilities at the SRS because of the additional jobs they would bring to the area. Both supporters and opponents of the proposed facility mentioned during interviews that elected officials are aware that not supporting such developments would lose them the support of Aiken County voters.

In terms of nuclear issues, the most vocal and active local group in the area is the Center for Nuclear Technology Awareness (CNTA) – a nonprofit, grassroots organization formed in 1991 to support the SRS site and be "pro-nuclear and proud of it" (Citizens for Nuclear Technology Awareness 2010). The CNTA representatives that we interviewed were quite proud of the reputation and influence of the organization at the local and national level. One interviewee said that if the mayor of Aiken wanted to meet with community leaders, the executive director of CNTA would be included among them. Leaders in the organization described being called by reporters at the *New York Times*, *Denver Post*, and *San Francisco Chronicle* for interviews and quotes, and editorials from board members have been published at these newspapers, as well as the *Aiken Standard* and *Augusta Chronicle*. CNTA organized a local campaign in support of the MOX facility, in which locals mailed postcards to the Department of Energy (DOE) calling for the MOX facility to be built at the SRS. An interviewee joked that the DOE received so much mail that they called "and asked [CNTA and the community] to stop sending postcards."

As in Claiborne County, "outside agitators" were the biggest sources of opposition to the MOX facility and organized most of the opposition activities that we captured in our scoring metric. They attended and spoke at public meetings on the EIS, wrote letters to and editorials for local and national newspapers, lobbied Congress to halt development plans, organized workshops and debates at the Aiken campus of the University of South Carolina, and organized two small protests against the shipment of plutonium into the SRS. Among the outside organizations that opposed the proposal were the Blue Ridge Environmental Defense League, the now defunct Nuclear Control Institute, the Movement for Nuclear Safety, the Carolina Peace Resource Center, the South Carolina Progressive Network, Georgians Against Nuclear Energy, Citizens for Environmental Justice, Citizens Against Plutonium, Greenpeace, and Friends of the Earth. All these organizations maintain offices outside Aiken County, mostly in Charleston, Columbia, and Charlotte, North Carolina. However, their activities had little influence in the community of Aiken. A former state employee said that it was difficult for these organizations to gain entrée into the Aiken community because they were viewed as outsiders who did not understand what the SRS had done for Aiken. He added that the siting decision was a federal decision, and, because regional groups had very little influence at the federal or local levels, their efforts at opposition were largely ineffective. This was echoed by a vocal opponent of the project, Tom Clements. Clements, who has represented the Nuclear Control Institute, Greenpeace, and Friends of the Earth, said that efforts to get locals in Aiken to ask hard questions of the DOE and the SRS rarely succeeded because of the work of "boosters" like the CNTA that were "in the pockets" of the DOE and the management companies because of "the money." In his opinion, local public interest about social issues paled beside "big government money." Clements went to great pains to point out that although he was based in Columbia, he was a local, having "roots in the area" because he had been born in Savannah.[30]

One issue did get raised in connection with the development of the facility: where to store the plutonium prior to the start of operations. Governor (1999–2003) Jim Hodges was concerned because plutonium from across the country would be brought into South Carolina to be stored at the SRS before the MOX fabrication facility had been approved. The issue was that no plans for the storage or removal of the plutonium – should the facility not be approved – had been made. There was much concern that South Carolina would become a dumping ground for nuclear waste if the MOX facility was never built. When asked by the press, Governor Hodges said that he would be willing to take all means possible to prevent this from happening, including, if necessary, lying down in the middle of the street to prevent the plutonium from being brought into his state. This earned him the nickname "Governor Speedbump." Despite this publicity hiccup, all the interviewees stressed that Hodges supported the development of the facility.

Thus, in many ways, apart from the involvement of outside groups, the response of the local community in Aiken County was very similar to the reactions in San Patricio County and Cameron Parish to those proposals. Local residents did not merely accept but actively campaigned for the MOX facility as a way to spur local economic development. Local mobilization was in support of the proposed facility.

---

[30] T. Clements, interview with R. Gong, August 4, 2009.

Brazoria County (Freeport LNG)

We now turn to an example of a different recipe for low levels of mobilization – similar industry, experience with a past proposal, and no economic hardship. In this case (and the other cases like it), a small level of localized resistance developed but not enough to categorize this response as mobilization. Given that the community had experience with a past proposal and no economic hardship, it is not surprising that some members of the community were not happy about the proposal. However, the existence of a similar industry is a strong deterrent of mobilization, as we will see in the case of the Freeport LNG proposal in Brazoria County, Texas.

The Freeport LNG was proposed on Quintana Island, which is located just across the ship channel from Freeport. Quintana Island is inhabited by about fifty residents, although the population can swell dramatically with visitors on weekends and holidays. Brazoria County has a long history of industrial development and is home to major petrochemical and refining facilities. The city of Freeport was originally founded by Freeport Sulphur Company (now Freeport-McMoRan) in 1912 to exploit what were then the world's largest sulfur mines. In 1940, Dow Chemical purchased eight hundred acres bordering Freeport Harbor to construct an $18 million plant to extract magnesium from water. It quickly became apparent that there was not enough housing in the area for the plant's workforce, so Dow purchased additional property to build the nearby residential city of Lake Jackson. Today, Dow Texas Operations in Freeport has become Dow's largest integrated site. It covers more than five thousand acres and seventy-five individual production plants that employ more than 4,500 people. It is the largest employer in Freeport and manufactures more than 46 percent of Dow products sold in the United States and 23 percent of Dow products sold globally (U.S. Department of Interior 2008). As a result, Dow and the petrochemical industry dominate the economics of the county. "The economic impact of [the petrochemical] industry in Brazoria County – including payroll, taxes, capital spending, donations and local purchases – totals $2 billion a year" (Shaw 2004). According to the local newspaper, the chemical industry (heavy users of natural gas) supports 40 percent of the tax base in Brazoria County (Smith 2005). As a large user of natural gas, Dow was supportive of the Freeport LNG proposal and signed a purchasing agreement for five hundred million cubic feet of gas per day upon completion of the facility (Antosh 2003).

In addition, there were a great deal of concerns at the time of the Freeport LNG proposal about industry moving overseas. Several articles in the newspaper lamented the future of the petrochemical industry and the need for cheap supplies of natural gas to ensure the continued existence of this industry in Brazoria County. Moreover, Port Freeport commissioners felt competition to site a facility before other areas in the state beat them to it. Upon signing a lease with Freeport LNG, one commissioner stated that, "Brownsville, the Corpus area and Sabine Pass were all being considered. The fact that we had other competition and were successful speaks well of what we had to offer them [Freeport LNG]" (as quoted in Baker 2002).

Given this kind of domination by a single industry and a single company, it is not surprising that county residents have little history of activism against industrial development. One longtime activist in Brazoria County, Sharron Stewart, lamented that citizens in the county just don't organize around environmental issues. Because of the dominance of industry and Dow, in particular, many residents are either supporters of the industry or too scared to speak up. Because so many people in the county are employed by these companies, they are often afraid to lose their jobs. She cited a few examples of times when community members have formed grassroots opposition groups to proposed industrial development in the county (e.g., Friends of the San Bernard River and Bastrop Bayou Watershed Protection Association). However, these groups tend to form along recreational rivers in unincorporated areas where a great deal of wealthy people from other places have weekend and vacation houses. In contrast, Quintana Island is a "blue collar beach" inhabited by yearlong residents.[31]

During the late 1990s and early 2000s, there was a heated debate about a proposal by the Texas Department of Transportation (TxDOT) to construct a bridge to Quintana Island. The island had previously been served by a swing bridge. When marine traffic approaches, which is quite often, the bridge deck simply floats down the channel or swings out of the way. Once the ship has passed, a system of cables and pulleys draws the bridge deck back. For years, the people of Quintana maintained a love-hate relationship with the swing bridge. Right-of-way was granted to sea traffic, so cars sometimes had to wait forty-five minutes or more for a chance to cross the channel. That inconvenience kept the island quiet and its population low. At the same time, it caused problems for emergency traffic. Ambulances, fire trucks, and police cruisers had to yield to passing ships. Supporters of the TxDOT proposal said a new bridge would be cheaper than the swing bridge, estimated to cost $400,000 per year to operate and maintain. But opponents said it was just a ploy by developers to get their

---

[31] S. Stewart, interview with second author, May 21, 2009.

hands on another pristine part of the Texas coast and that money for the $9 million bridge would be better spent on other projects like upgrading the area's hurricane evacuation road.

Teresa Cornelison, a relatively new resident to Quintana Island at the time, led the battle against the new bridge and would later oppose Freeport LNG. The town of Quintana signed a resolution against the bridge's construction. TxDOT ignored the community's opposition and approved the bridge. The swing bridge stopped operating in March 2003 when the TxDOT bridge was completed. Some residents thought TxDOT's enthusiasm for the new bridge was linked to the LNG proposal, a proposal that would have been incompatible with the original swing bridge. The case of the bridge provides the lone example of Quintana Island residents' experience in opposing a previous development proposal.

Although several Quintana residents, including Cornelison, were strident opponents of the facility and eventually partnered with birding groups from nearby Lake Jackson and Houston, Dow's support made it difficult for this opposition to spread elsewhere in the region. Local opposition was quickly dismissed by other residents: "The Texas Gulf coast has been in the center of the petrochemical industry for generations. To say that the area can't accommodate a LNG operation would be like choking on a gnat after swallowing a camel" (Hawes 2004). At the Draft EIS hearing, several residents, including then Mayor James Nevil, made similar comments:

I can understand the concerns of the people that live in Quintana, in fact all of us in this area, of what might happen [if the LNG facility comes]. But at the present time, Quintana is right across the waterway from Dow Chemical Company which has hundreds of tanks and thousands of other vessels with flammables in them . . . . Coming in the jetties up and down the Intercoastal Canal, there's hundreds of ships, barges, containing flammable materials which have been regulated to the point where they have not been a big problem. There's a big tank of ammonia that belongs to BASF not very far from Quintana. There's a big discharge point for ships of oil that comes in from offshore that is very close to them . . . I think we have enough regulatory bodies already that can see that these things are done safely, that whatever concerns about the birds having to move a few hundred yards away can be taken care of, and I don't see any reason why we should not put a facility here where it is needed for the energy needs of this area. (V. L. Scott as quoted in Federal Energy Regulatory Commission 2003, 10–11)

Some on Quintana, don't like the idea of a tank farm in their backyard, but we are in the backyard of industry. We have probably the biggest chemical plant on the other side of the canal, we have the canal, the harbor, the strategic oil reserve right down the road, it's in our backyard already. (J. Nevil as quoted in Federal Energy Regulatory Commission 2003, 40)

As a result, opponents of the Freeport LNG facility focused efforts on securing mitigation.[32] These efforts resulted in the relocation of a popular birding park on Quintana Island and the purchase of seventy-eight acres in nearby Surfside for wetlands mitigation.

The Freeport LNG facility was approved and built, but relations between the company and community remain strained. In Quintana, a battle is currently being waged between local residents and Freeport LNG over the company's recent request to truck LNG to the facility.[33]

CONCLUSION

We have covered a great deal of ground in this chapter. We close by summarizing the results reported here and with a bit of speculation concerning differences between the "rights struggles" that have dominated scholarship in the field and the kind of local conflicts over technical, land-use issues with which our research is concerned. We begin with the summaries. In this chapter we have sought to answer three questions. First, how much opposition mobilization did we actually find in our twenty communities? Second, what factors and dynamic mechanisms account for variation in the mobilization we observe? Finally, how are we to understand the cases that featured no mobilization? What factors/mechanisms predict these cases?

We can dispense with the first of these questions fairly easily. We found remarkably little emergent opposition across our cases. By the operational criteria we employed in the project, half of our communities experienced "mobilization." But the level of mobilized activity was exceedingly modest. Across all twenty communities we saw a total of just twenty-seven protest events and three lawsuits. Nor was there all that much more routine, institutionalized response to the proposed projects. For example, the median number of letters to the editor concerning the proposals was just 5.5 across our cases. Even we were surprised by the paucity of opposition activity in the communities. If we are now living in a "movement society," apparently many of our fellow citizens have yet to get the memo.

The results that bear on the two remaining questions merit a bit more discussion. We won't bore you with another detailed recounting of the various recipes reported for the two outcome conditions: mobilization and nonmobilization. Instead, we summarize what we see as the central thrust

---

[32] Ibid.
[33] W. Neeley, interview with second author, May 22, 2009.

of those recipes. With regard to mobilization, we found that political opportunities and civic capacity explained higher levels of mobilization, but that these staples of social movement research tended to work in combination with – and probably through – our community context variables. These include such causal conditions as similar industry, economic hardship, and experience with a past proposal. We interpret the former two causal conditions as reflecting the objective structural propensity for collective action within the community and the latter cluster of context variables as powerfully shaping the subjective interpretation of project risk by local residents. Or to simplify things even further, we see opportunity and capacity as determining the objective potential for contention and the context variables as shaping the motivation to act on that potential.[34] In this sense, we see the results as mirroring the conjunctural logic and substantive thrust of the original formulation of political process theory (McAdam 1999 [1982]). In his initial sketch of the perspective, McAdam stressed the critical importance of emergent, constructed assessments of threat and opportunity (e.g., cognitive liberation) as the key catalyst of mobilization, with improving political conditions and the presence of "mobilizing structures" only affording insurgents the capacity to act but not the motivation to do so.

We interpret our nonmobilization recipes in much the same way. Again, we find that our two standard social movement variables – civic capacity and political opportunity – figure prominently in the recipes explaining nonmobilization, though it is their absence, rather than presence, that bears on the outcome condition. Once again, however, our results improve substantially when we add the community context variables to the recipes. Designed to capture the perceived risks posed by a project, these context variables represent a stand-in for the stress on cognitive liberation in the original formulation of the political process model.

As promised, we close on a more speculative note. Although we have emphasized the continuity between our findings and the initial formulation of the political process model, we are nonetheless struck by an essential difference between "rights struggles" such as the civil rights movement – from which political process theory was derived – and the very different kind of conflicts that we are interrogating here. The difference concerns the causal salience of "threat" versus "opportunity" in the two types of

---

[34] Our stress on the critical importance of the community context and the interpretation of risk by local populations is very much in accord with similar findings reported by Coppens (2011) and Sherman (2011).

struggles. The concept of political opportunity has been seen as critically important to the emergence of the kind of rights struggles that have been the focus of so much movement scholarship. Given the nature of these rights movements, the emphasis on perceived opportunities makes a great deal of sense to us. Normally, in such cases, the nature of the central issue – group-based discrimination and disadvantage – is clear, as is the underlying motivation to remedy it. It is only the long-standing power imbalance between the aggrieved population and its opponents that forestalls mobilization. In such cases, movement emergence ordinarily does require some kind of rupture or crack in the system – an opportunity of some kind – as a catalyst to action.

The kinds of disputes that concern us here, however, are quite different from the aforementioned rights struggles. By whatever term we refer to them – siting controversies, NIMBY movements, or locally unwanted land-use disputes (LULUs) – the issues that ultimately motivate these disputes are initially far more ambiguous and uncertain than is true for rights movements. In the latter, the issues are generally clear enough, even if the outcome of the struggle remains highly uncertain. Siting disputes involve issues – oftentimes highly technical in nature – that are not clear on their face. This uncertainty, as Tversky and Kahneman (1974) noted years ago, tends to stimulate collective efforts at sense making and social construction. But unlike the rights struggles, these efforts are far more likely to turn not on perceptions of opportunity but on socially constructed conceptions of threat. The implication we draw from this is a simple one. Instead of assuming that one theoretical model fits all movements, we should be attentive to systematic differences in the character and dynamics of different categories of movements and modify our theories accordingly.

4

Does Opposition Matter?

Mobilization and Project Outcome

In the previous two chapters we showed that there was considerable variation in mobilization across our twenty communities. In this chapter we turn
our attention to the all-important issue of "movement outcomes." At the
heart of the chapter is a single, stark, research question: does the aforementioned variation in mobilization help us understand the ultimate fate of the
proposed projects? To answer the question, we look at three different "outcome" measures. The first is whether the project was rejected during the
regulatory review process. The second is the opposite of the first, whether
the project was approved during the regulatory review process. The third – a
very different outcome – is whether the project was ultimately built.

Given the exceedingly modest levels of contention that we found across
our cases, we were frankly expecting to see little affect of mobilization/
nonmobilization on project outcomes. Taken together, however, our results
show a surprisingly strong imprint of mobilization/nonmobilization on the
ultimate fate of the projects. Three specific findings bear mention. The
strongest relationship is a negative one: the failure of a community to
mobilize is, in and of itself, sufficient to explain project approval.
Although the converse is not true, mobilization *is* an important component
of the recipes explaining project rejection. The outcome that shows the
weakest relationship to mobilization/nonmobilization is whether the project
is ultimately built. This makes sense. The decision to build a project depends
on a host of factors – most importantly in our cases, the variable health of
the relevant energy market at the time of construction – other than level of
opposition. Still we will include all three measures in our analysis. But we are
getting ahead of ourselves. Before we dig into the data, we want to review
the curious history of research – or lack thereof – into this seemingly most
important of research questions.

RESEARCH ON MOVEMENT OUTCOMES: WHY SO LITTLE, SO LATE?

The ultimate justification for studying social movements would seem to be obvious. Those who are drawn to this form of action certainly assume that movements are potentially significant vehicles of social and political change. But are they? If so, what are the specific mechanisms that mediate their impact? As logical and straightforward as these questions are, however, systematic research designed to answer them was late in coming and remains relatively sparse. Some twenty years after the field of social movement studies had emerged, McAdam and Snow (1997: 461) could still lament "that the impact of social movements has been the subject of so little systematic scholarship." Since then, a body of work on the topic has begun to develop, but it is still relatively thin and plagued by a host of empirical problems that often make it difficult to reach any firm conclusion about the extent of a movement's impact in any given case. We will return to this issue in the following text. But we first want to take up the logically prior question of why it took movement scholars so long to take up the issue of "outcomes."

To answer the question we return to the "sociology of knowledge" history of the field sketched in Chapter 1. Reflecting that history, the answer to the question seems relatively straightforward to us. Movement scholars did not assign much importance to the question of outcomes because they believed they already knew the answer. This applied as much to the classic collective behavior theorists as to the new generation of scholars who birthed the field of social movement studies during the 1970s and early 1980s. We have already summarized the view of collective behavior theorists on the matter in Chapter 1. Having defined social movements – and all forms of collective behavior – as an ineffectual collective response to a social system under strain, it would have been odd to invest serious time and energy in conducting studies of their substantive political and policy impact. At best, movements function as a precursor to effective political action by alerting rational policy makers to sources of strain in society to which they will need to attend (Smelser 1962: 73). But, in and of themselves, movements function at a psychological rather than a political level and should be studied accordingly. In keeping with this view, the central thrust of work in this area was on the psychological and social psychological origins, dynamics, and functions of social movements (Feuer 1969; Hoffer 1951; Klapp 1969; Lang and Lang 1961; Smelser 1962).

As recounted in Chapter 1, the social-political turbulence of the 1960s and early 1970s greatly increased scholarly interest in social movements and led to a conceptual revolution in the field. The resulting paradigm shift was championed by those whose own experience convinced them, not only of the importance of movements but also of the need to jettison the collective behavior framework in favor of perspectives that stressed the link between movements and more routine forms of political and organizational life. The development of these alternative perspectives – resource mobilization and political process in the United States and new social movement theory in Europe – fueled the rise of what is now the very large and well-established interdisciplinary field of social movement studies. But we are getting side-tracked. How does all of this relate to the issue of movement outcomes?

Having reclaimed the study of social movements as the proper province of political analysts and organizational scholars, you would have thought that systematic studies of their substantive political and policy impact would have become a centerpiece of the new field. But such studies were not forthcoming. We think the reason is simple and reflects the generally partisan, movement-centric views of those who shaped the field at its outset. Although collective behavior theorists assumed the general ineffectiveness of movements, the new generation of scholars shared the opposite conviction. Having taken part in, or strongly identified with, the movements of the New Left, these pioneering scholars were already convinced of the singular importance of movements as catalysts of social and political change. Instead, the central analytic focus of the field quickly coalesced around the question of movement emergence (McAdam 1995) and the dynamics of mobilization (Walder 2009a). Having assumed the causal potency of movements, the central issue shifted to understanding the conditions under which movements developed.[1]

---

[1] One might reasonably ask, where was American political science as the social movement subfield was developing within sociology? With a few notable exceptions (e.g., Eisinger 1973; Lipsky 1968; and Tarrow 1983), the conventional view in political science has been one that attributes little influence to social movements as vehicles of political or policy change. Some of this reflects the discipline's central empirical preoccupation with formal political institutions (e.g., Congress, the presidency, and the courts). If social movement scholars have fashioned a Ptolemaic view of politics by locating movements at the center of their cosmos, political scientists are similarly guilty of propounding a distorted view of politics by essentially treating movements as part of the background "noise" or "error term" of political life.

But other emphases within political science have reinforced the generally dismissive view of social movements. At least two dominant lines of work in the field have converged to support this view. The first is the voluminous literature spawned by Olson's 1965 book, *The*

By now it should be clear how we would regard the assumption that movements are ordinarily a consequential force for social-political change. When combined with the central empirical focus on movements – and relative neglect of other political actors – this assumption tends to reinforce all of the Ptolemaic tendencies touched on in earlier chapters. By making movements the central, if not exclusive, object of research attention and simply assuming their effects, scholars are once again virtually assured of "confirming" the substantive importance of the phenomenon. We obviously favor a different approach. It is no more legitimate for today's social movement scholars to assume the political impact of movements than it was for proponents of the collective behavior perspective to dismiss movements as irrational and ineffective. Rather the matter should be among the central empirical issues of concern not only to social movement scholars but also to analysts who are intent on understanding the factors that shape the outcome of legislative and other policy processes.

The good news, even if it was late in coming, is that movement scholars have, in recent years, begun to turn their attention to this issue (Amenta 2006; Amenta, Olasky, and Carren 2005; Andrews 1997, 2001, 2004; Burstein 1993, 1998; Cress and Snow 2000; Earl 2000; Giugni 1998; Giugni et al. 1999; McAdam and Su 2002; McCammon et al. 2001, 2007; Soule et al. 1999). This literature is not without its flaws, however. We worry, in particular, about two sources of bias that characterize much of the research in this area. The first of these biases owes to the by now familiar problem of selecting on the dependent variable. Quite simply, by

---

Logic of Collective Action. Concluding that it is irrational for someone to engage in collective action in pursuit of "public goods" when they cannot be denied the benefits of such action, Olson and his many disciples have fashioned a powerful perspective that strongly implies the ineffectiveness of movements as a force for policy change. There are actually two distinct implications here. The first concerns the possibility of mounting effective collective action at all. If rational actors refrain from such action, then it should be nearly impossible to organize a movement in the first place. The second implication has to do with those who might take part in a movement, given its demonstrated irrationality. If rational actors can be counted on to eschew participation, then only those with nonrational motives are likely to gravitate to social movements, and this will generally keep such movements small and politically inert.

The latter implication shades into the second body of work alluded to in the preceding text. This is the rich, if amorphous, work in political science that stresses the strategic preference of elected policy makers for broad, centrist policies that can attract majority support. Down's (1957) work on the "median voter" is emblematic and influential in this regard. When coupled with research showing that movement activists are typically far more ideological and extreme in their policy preferences than the population at large (Finkel, Muller, and Dieter-Opp 1989; Nie and Verba 1975), the implication is clear: their characteristic unrepresentativeness should render social movements inconsequential as a political force.

picking especially large and celebrated struggles for study, social movement scholars would appear to have stacked the deck when it comes to offering any kind of assessment of the social change potential of social movements. Our point here is but an extension of the argument we made in Chapter 1. If it seems problematic to try to understand the dynamics of emergent collective action by studying only those exceedingly rare instances of widespread and prolonged mobilization, how much more problematic it is to regard the outcomes of such struggles as being at all typical of the social change impact of movements. Among the most interesting questions for us in this study is whether much more modest, and presumably typical, levels of mobilization have any impact on the fate of the projects they are designed to address.

The second problem that characterizes much of the work on movement outcomes stems from the more general challenge inherent in trying to empirically account for any complicated macrolevel social change process. Perhaps an example will help to illustrate the problem. Take the case of the women's movement in the United States. One could easily generate a list of changes in American society since, say, 1975 that could plausibly be attributed to the women's movement. These would include the expansion of abortion rights occasioned by *Roe v. Wade*, the increase in the number of female elected officials, higher rates of conviction and stiffer sentences for rape and violence against women, the spread of gender-neutral language, and the entrance of large numbers of women into previously male-dominated occupational domains (e.g., law, medicine, and science). But generating such a list is hardly the same as proving the connection between the movement and any of these changes. To move beyond the realm of the plausible, the researcher must confront the hoary issue of spuriousness. A spurious relationship occurs when two phenomena are themselves the product of a third phenomenon that is causally prior to the other two. Returning to our example, several prior social processes have been proposed as the primary "cause" of the women's movement and the more specific changes listed in the preceding text. For illustrative purposes we will confine ourselves to just one of these alternative arguments.

John Meyer and various colleagues contend that throughout the past century or so, but especially after World War II, we have seen the spread of Western "civic" norms around the globe (Boli and Thomas 1997; Meyer, Boli et al. 1997; Meyer, Frank et al. 1997). As a result of this generalized diffusion process, policies that emphasize human rights, environmentalism, and, most relevant to our example, women's rights are now among the central normative "requirements" for standing among nations in the current global order. From this perspective, the women's movements we see around the globe – including in the United States – are more an expression of the underlying diffusion process than a force for change in their own right. We might counter by asking where these civic "norms" came from if not from the feminist movements that developed in the democratic West during the 1960s and 1970s. We cannot hope to resolve the issue here; we raise it simply to underscore just how difficult it is to prove the causal power of movements relative to other broad change processes.

Even more than the matter of spuriousness, this is the real issue that makes attributing causal force to movements so difficult. That is, the discussion is pitched at such a broad, macrolevel that the connections between hosts of simultaneous change processes are simply impossible to disentangle. We return to our example one last time in order to underscore this point. Consider the following broad change processes that were occurring at more or less the same time in the post–World War II United States:

- The increase in women's participation in the workforce,
- Rising divorce rates,
- The growing availability of female contraception,
- A quarter of a century (1945–70) of unprecedented economic prosperity,
- The rapid expansion of female college attendance,
- The aforementioned diffusion of Western civic norms, and
- The emergence and subsequent development of the modern U.S. women's movement.

It is very likely that all of these change processes contributed to the gender revolution that took place in the United States during the late twentieth century. But attributing any of the outcomes listed in the preceding text to the "women's movement" seems like an exercise in futility. Yet at least some of the work on movement outcomes has very much this kind of amorphous feel. Some broad national struggle is plausibly linked to a set of specific outcomes but without any attempt to assess the causal contributions of a host of other contemporaneous change processes simultaneously. The real problem is not the causal attribution but simply the impossibility of trying to sort out influences at such an abstract macrolevel.

By contrast, the most convincing studies of movement impact are ones that frame the problem in more narrow geographic terms. Our model in this regard is Kenneth Andrews's (1997, 2001, 2004) careful study of the variable impact of the civil rights movement across all counties in Mississippi. By systematically measuring the level of movement activity and a number of electoral, antipoverty, and educational outcomes at the county level, Andrews was able to overcome many of the empirical problems inherent in more macrolevel studies of movement effects. In the end, Andrews was able to demonstrate a very close empirical connection between "sustained" movement activity and his various electoral and antipoverty outcomes. By contrast, the data showed that his measure of educational desegregation (e.g., the index of dissimilarity) was impervious to movement influence. Just as important, Andrews's deep knowledge of the movement in these various locales allowed him to offer convincing accounts of the underlying social processes that explained his positive and negative results. For example, he was able to show that counties with the highest levels of movement activity also experienced the most dramatic growth in private, white, educational "academies" during the 1970s and 1980s. By allowing for "white flight" from the public school system, the private academy "countermovement" enabled segregationists to undercut the movement's efforts in this particular institutional realm.

In seeking to assess the impact of local mobilization on the ultimate fate of the proposed projects, we have taken a page from Andrews's book. By moving from the county level down to the local community level, we have sought to go Andrews one better. We are in an even better position, we think, to assess systematically, net of other causal conditions, the link between the level of local opposition and project outcome. Moreover, by selecting twenty cases at random, we are in a very good position to see whether much more typical, low levels of contentious politics have any influence on the outcome of these projects.

RESULTS

We noted at the outset of this chapter that we are going to organize our analysis around three outcomes: approved, rejected, and built. But these outcomes are but part of a more complicated continuum. We can distinguish between five outcome "states." These are

- Project rejected,
- Project withdrawn,
- Approved but withdrawn,
- Approved but postponed, and
- Project built.

Adding these other three outcomes to the two main ones gives us a continuum that ranges from rejected, as the most favorable resolution the opposition could hope for, to built, at the other extreme. The full continuum is shown in Figure 4.1 along the *x*-axis, with the *y*-axis depicting the range of fuzzy-set values used to measure the outcome condition: opposition mobilization.

All of our cases are then arrayed in the interior of the figure at the intersection of their values on the outcome and mobilization scales. We begin by presenting our data in this way to give the reader a general sense of the relationship between mobilization and outcome. Given the relatively low levels of mobilization discussed in Chapter 2, the general relationship shown in the figure may come as something of a surprise. Despite the low levels of opposition activity, there is little question but that higher levels of mobilization are associated with more favorable project outcomes, from the opposition's point of view. That is, most of the cases with higher levels of mobilization are clustered in Figure 4.1 in the lower right-hand quadrant, reflecting troubles for the project (e.g., rejection or withdrawn) and a victory of sorts for the opposition. Expressed another way, as of this writing, only one of the ten projects with mobilization values of 0.6 or higher have been built, while seven others have been either rejected or withdrawn. Conversely, nine of the ten with mobilization values of 0.4 or less have been approved. Even comparatively low levels of emergent collective action would appear to yield less favorable project outcomes. The reverse is true as well. The absence of contention is clearly related to more favorable project outcomes as shown by the clustering of the latter in the upper left-hand quadrant of the figure. But we will want to explore these relationships in more depth using the same combination of fs/QCA and narrative analysis employed in Chapter 3.

To do so we begin by recoding these different outcome states into the three outcome conditions – approval, built, and rejection – that will be the focal point of our analyses. So, for example, to score rejection, all projects that were at any time rejected by decision makers scored a 1, projects that were withdrawn in the face of imminent rejection scored a 0.8, projects that were withdrawn for reasons not clearly related to imminent rejection score a 0.5, and all other projects score a 0. Approval is scored as the opposite of rejection. That is, all projects that were approved – even if they were subsequently postponed or even withdrawn – were scored a 1, projects that were withdrawn under ambiguous circumstances scored a 0.5, those that were withdrawn in the face of rejection received a 0.2, and finally those that were rejected received a 0. Finally, built is scored as 1 for projects that are currently either under construction or have already been

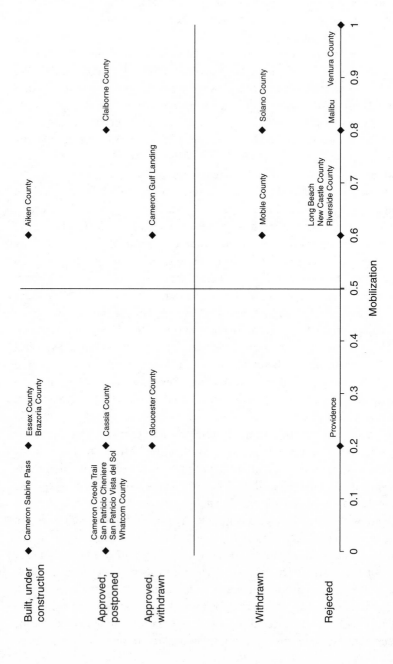

FIGURE 4.1. Opposition and project outcome

TABLE 4.1. *Project Outcome Fuzzy-Set Scoring*

Project Name	Project Outcome	Fuzzy-Set Scoring		
		Approved	Rejected	Built
Cameron Sabine Pass	Built	1	0	1
Brazoria	Built	1	0	1
Essex	Built	1	0	1
Aiken	Built	1	0	1
San Patricio Cheniere	Approved but postponed	1	0	0
San Patricio Vista del Sol	Approved but postponed	1	0	0
Cameron Creole Trail	Approved but postponed	1	0	0
Whatcom	Approved but postponed	1	0	0
Cassia	Approved but postponed	1	0	0
Claiborne	Approved but postponed	1	0	0
Gloucester	Approved and withdrawn	1	0	0
Cameron Gulf Landing	Approved and withdrawn	1	0	0
Solano	Withdrawn before approval	0.5	0.5	0
Mobile	Withdrawn before approval (likely rejection)	0.2	0.8	0
Providence	Rejected	0	1	0
Long Beach	Rejected	0	1	0
Riverside	Rejected	0	1	0
New Castle	Rejected	0	1	0
Malibu	Rejected	0	1	0
Ventura	Rejected	0	1	0

built, and built is scored 0 otherwise. The full set of outcome scores is shown in Table 4.1.

We proceed as we did in Chapter 3, by seeking first to explain each outcome condition through the technique of fuzzy set/Qualitative Comparative Analysis (fs/QCA) and then illustrating those results through analytic narratives of one or more cases. We take up approval first.

Nonmobilization and Approval

The strongest relationship that we found between opposition and project outcome is that between nonmobilization (or the lack of mobilization) and approval. Lack of mobilization is sufficient by itself to explain approval, at a consistency of 0.80. A lack of mobilization is also almost a necessary condition for approval, at a consistency of 0.76.[2] This finding makes theoretical and intuitive sense. In locations where there is little opposition to a project, it is easily approved by regulators. Of the ten cases in our sample that exhibit little mobilization, only one was not approved – KeySpan LNG in Providence County, Rhode Island. A thorough examination of this case (presented in the section on rejection in the following text) provides evidence of how external actors, specifically state officials and opponents of a nearby, similar project, can sometimes overcome the lack of local mobilization and derail a project. In this sense, we regard the Rhode Island case as the exception that proves the rule. In and of itself, the absence of local opposition is generally enough to ensure project approval. The clarity of this result speaks to the wisdom of studying cases other than ones of successful mobilization. It also provides another interesting perspective on the efficacy of contentious politics. Analysts have sought to demonstrate the influence of movements exclusively by showing that successful mobilization yields outcomes preferred by activists. But one can, as we have, ask the opposite question: is the failure to mobilize linked to unfavorable outcomes? In our case, the answer is clearly yes. A detailed account of BP's Cherry Point proposal for a cogeneration project in Whatcom County, Washington, will further illustrate this relationship.

Whatcom County, Washington (BP Cherry Point)

In June 2002, BP first proposed a cogeneration facility to provide steam and 85 megawatts of electricity to meet the operating needs of its adjacent refinery and 635 megawatts of electrical power for local and regional consumption. The proposed facility was to be located between Ferndale and Blaine in northwestern Whatcom County, Washington, about eight miles from the Canadian border, on BP-owned, unimproved property, zoned for heavy impact industrial development. The existing BP Cherry Point refinery has been operational for more than thirty years, processing 225,000 barrels of Alaska North Slope crude oil a day in order to provide about 20 percent of the gasoline market share of Washington and Oregon;

---

[2] Consistency scores of greater than 0.8 for sufficient conditions (or causal combinations) and 0.9 for necessary conditions are commonly used as guidelines by scholars of fs/QCA in an effort to establish the relevant set-theoretic relationship between the causal and outcome conditions.

provide the majority of jet fuel to Seattle, Portland, and Vancouver, B. C., international airports; and serve as the largest West Coast supplier of jet fuel to the U.S. military. In addition, the refinery employs about 825 full-time workers. Thus the refinery serves an important role locally by providing jobs and tax revenue and regionally by providing necessary transportation fuel. Moreover, BP had developed a reputation in the community as a good corporate neighbor through frequent charitable contributions and a safe operations record. BP argued that the new boilers and technology associated with the proposed cogeneration facility would actually reduce air pollution from the existing refinery and water withdrawal from the Nooksack River.

Several contextual issues structured the warm reception granted by the local community to this proposal. Soaring energy prices and rolling brownouts and blackouts (associated with electricity deregulation in California and the Enron scandal) were major issues for Whatcom County in 2001 and had resulted in the closure of several local industrial plants. Energy supply was such a problem that one legislator suggested building a new government-owned and -operated power plant in Cherry Point. Debate about the project also came on the heels of the September 11th attacks. BP's Michael Abendoff noted during our interview that the tone surrounding discussions of energy supply focused on energy security and independence during this time.[3]

The community was well aware of the risks associated with energy facility development and well equipped to mobilize against such proposals. The nearby city of Bellingham had been the site of a devastating pipeline explosion in 1999, which killed three local boys. This incident has had a major impact on the community and resulted in the creation of the Pipeline Safety Trust – an organization dedicated to promoting fuel transportation safety through education and advocacy. Moreover, a similar proposal in 1999 for a natural gas-fired generation facility, Sumas Energy 2 (SE2), had generated a significant amount of opposition in Canada and the United States, with hundreds attending hearings and writing comment letters. After initial rejections by Washington State's Energy Facility Site Evaluation Council (EFSEC), SE2 submitted revised applications in 2000 and 2001, and the governor of Washington State approved the facility in August 2004. However, citing concerns about potential air pollution emitted from the facility, Canadian officials denied the permit to construct the associated transmission line to a Canadian substation in 2004 – a

[3] M. Abendhoff, interview with R. Wright, July 22, 2009.

denial that the Supreme Court of Canada upheld in early 2006. The facility was withdrawn later that year. Thus the community surrounding the facility was well versed in techniques for opposing such facilities and knew how to make its voices heard.

However, for a number of reasons, the community chose not to oppose the BP Cherry Point Cogeneration proposal. One environmental leader noted that the project was not a big priority to her organization because it was located in an area that was already industrial and represented a "reasonable alternative" to current uses.[4] Another environmentalist, whose home was just steps from BP property, revealed that the small amount of opposition that did develop against the project was more about maximizing mitigation than disagreements about the facility.[5] State Representative Doug Erickson summed up local attitudes toward the project best when he explained why he commented in favor of the proposal at an EIS hearing, "BP is one of our best employers ... a huge tax generator, a great facility ... they have been and continue to be excellent corporate citizens and neighbors here in Whatcom County so how can you not support them when they're doing so many good things for the community?"[6] Though concerns about wetland mitigation extended the time to final approval, the proposal was approved in December 2004. However, BP has yet to build the project, claiming that company strategy and focus has changed since it was first proposed. We will have more to say about the relationship between lack of mobilization and construction later.

Nonmobilization and Built

Our data show a slightly different but still important relationship between nonmobilization and construction. Unlike approval, the absence of mobilization is not sufficient in and of itself to explain whether a project gets built. A lack of mobilization is close to being a necessary condition for built, with a consistency of 0.75. Again, this finding makes theoretical and intuitive sense. If, as we saw in the previous section, nonmobilization is sufficient to explain approval, and approval is, in turn, required for a project to be built, then it follows that nonmobilization is also necessary

---

[4] W. Steffensen, interview with R. Wright, July 23, 2009.
[5] E. Friloeb, interview with R. Wright, July 23, 2009.
[6] D. Erickson, interview with R. Wright, July 22, 2009.

to get a project built. It is not, however, sufficient, as the KeySpan case described in the following text will make clear.

There are myriad reasons why a project might not get built that have nothing to do with mobilization or the lack thereof. Chief among these are economic considerations, none more important than the state of the relevant energy market at the time construction is slated to begin. Other kinds of market dynamics can be relevant too. Consider the case of the Idaho wind farm described in Chapter 3. Although approved at the close of the review process and at a time when demand for energy was unusually high, the sponsoring company, Windland, was nonetheless unable to attract buyers for the energy expected to come from the project. In the end, the lack of adequate transmission lines linking the site to the broader energy grid in the Northwest was blamed for the lack of subscribers. When confronted with these "negative market externalities," it simply made no sense to build the approved facility. Clearly, in such cases, nonmobilization is largely irrelevant to the decision to build. Of the twelve approved cases in our sample, only four have actually been built. The rest have either been withdrawn or postponed.

In order to test some of these ideas about other factors that may be relevant to project construction beyond mobilization, we again turn to fs/QCA. For the purposes of this exploratory analysis, we choose to focus only on liquefied natural gas (LNG) projects – to ensure that we have a clear grasp of market dynamics – and only those that were approved – an obvious necessary condition for project construction. As a result, our analysis is limited to ten of our original twenty cases (see Table 4.2). In addition to mobilization, we include two other causal conditions related to whether a project gets built: regional saturation and "irons in the fire."

Regional saturation is important, particularly in the case of LNG, because market demand for natural gas would support only eight to ten projects,[7] thus first movers were at a distinct advantage. To measure saturation, we looked at the number of projects approved and under construction at the time of the project's approval in the project's region (East, West, or Gulf) since the company's announcement of the project. Specifically, a case received a score of 1 if three projects were approved and under construction in the region by other companies since the announcement of the proposal, a score of 0.67 if two projects were approved and under construction, a score of 0.33 if only one project was approved

---

[7] G. Sweetnam, presentation at Department of Energy LNG Forum, June 1, 2006, Los Angeles, CA.

Putting Social Movements in Their Place

TABLE 4.2. *Fuzzy-Set Scores for Conditions in Built Analysis*

Case	Regional Saturation	Irons in the Fire	Mobilization	Built
Brazoria	0	0	0.2	1
Cameron Creole Trail	1	1	0	0
Cameron Gulf Landing	1	0	0.6	0
Cameron Sabine Pass	0.33	0.4	0	1
Essex	0	0	0.2	1
Gloucester	0.67	0	0.2	0
Mobile	1	0.4	0.6	0
San Patricio Cheniere	0.33	1	0	0
San Patricio Vista del Sol	1	0	0	0
Solano	0	0	0.8	0

and under construction, and a score of 0 if no projects were approved and under construction. Thus lower scores on regional saturation should be associated with projects getting built.

What we are calling "irons in the fire" is a proxy for the company's willingness to pursue a project beyond approval. Many companies, particularly in the Gulf Coast region, proposed projects in multiple locations. However, the intent of these companies was not to build in multiple locations but to see which locations moved through the approval process fastest. Thus whichever projects were approved first were the ones that got built. In order to capture this phenomenon, we measured the number of projects by the same company approved in the region at the time of the given project's approval since the company's announcement of the project. Specifically, a case received a score of 1 if more than one other project by the same company had been approved in the region since the company's announcement of the project, a score of 0.4 if only one other project by the same company had been approved, and a score of 0 if no other projects by the same company had been approved in the region. As with regional saturation, lower scores on the irons in the fire condition should be associated with projects getting built. Table 4.2 shows the fuzzy-set scores for all conditions included in the subsequent analysis; Appendix B provides the relevant information behind each of the scores.

Our results suggest that all three of these factors in combination (low mobilization, little regional saturation, and few irons in the fire) are important to explain when a project gets built, with a consistency of 0.81

and coverage of 0.73. All three cases in which projects were built – Brazoria, Essex, and Sabine Pass – fit this recipe. It is no coincidence that these three projects were among the first to be approved in their respective regions, highlighting the importance of first-mover effects and the potential importance of delays created by opposition efforts. These results demonstrate the importance of other factors beyond mobilization (or in this instance the lack thereof) in determining project outcomes.

We think our result would have been even stronger had we been able to include a nuanced measure of natural gas futures pricing among our causal conditions. However, we ran into several issues in trying to create such a measure. First, throughout the entire period of our study, natural gas prices were well above the threshold for profitability, that is, greater than $2.50 to $3.50 per million cubic feet.[8] Moreover, decreases in natural gas futures prices – owing to production from unconventional sources – did not occur until the end of 2006, affecting only the last three of our LNG projects. In addition, the models for predicting natural gas futures prices are notoriously inaccurate and therefore subject to wide variation. To render our measure more accurate, we had also hoped to incorporate information on the costs of the companies' supplying natural gas by using different technologies (e.g., domestic shale, foreign LNG, or domestic conventional). Knowing the mix of these sources is important to gauge a company's price sensitivity accurately, but data collection on this issue proved too difficult. For these reasons, we choose to focus on the three factors discussed in the preceding text. We would, however, encourage future researchers to try to build pricing conditions into their models.

To illustrate the three factor recipe described earlier, we focus on the two LNG proposals in San Patricio County, Texas. Together they nicely illustrate how approved projects fail to get built despite little mobilization, and instead succumb to market factors.

San Patricio County, Texas (Cheniere Corpus Christi and Vista del Sol)

These two proposed LNG facilities were to be located in an industrial area of Corpus Christi Bay, across from the city of Corpus Christi in nearby San Patricio County, Texas. The Port of Corpus Christi, which is the sixth-largest U.S. port and deepest inshore port on the Gulf of Mexico, handles mostly oil and agricultural products. Much of the

---

[8] D. E. Dismukes, presentation at Department of Energy LNG Forum, June 1, 2006, Los Angeles, CA.

local economy is driven by tourism and the oil and petrochemicals industry. Moreover, San Patricio County has a long history of oil and gas production, dating back to the 1920s. However, since the 1970s, this production has been in decline. At the time of the proposal, Corpus Christi was home to several military installations that the Pentagon was threatening to close. Several newspaper articles quoted local elected officials as supportive of LNG as a way to replace the revenue and jobs that would be lost due to closure of the bases. In addition, the area is home to several industrial plants (Sherwin Alumina, Elementis Chromium, Oxychem, and Air Liquide America Corporation) that use natural gas in their operations. When gas prices doubled in 2003, these companies were all adversely affected and seeking out alternative supplies of natural gas. Thus the surrounding community in San Patricio County and Corpus Christi was familiar with the oil and gas industry and actively seeking out new economic opportunities, specifically natural gas sources.

Cheniere Energy was the first to propose an LNG facility in the area during May 2003. The company had identified about five sites in the Corpus Christi area that met the technical criteria necessary for an LNG facility. They then took these potential sites to the presidents of local environmental organizations, specifically Coastal Bend Bays and Estuaries and the local Sierra Club chapter, and vetted them in terms of environmental criteria. The local environmentalists had issues with every site except the one that Cheniere selected for Corpus Christi LNG. This process occurred before Cheniere leased property in the area and a year before they submitted an application to the Federal Energy Regulatory Commission (FERC). It also gained the trust of the local environmental community and ensured their support of the Corpus Christi LNG proposal.

ExxonMobil soon followed suit in October 2003 with a second proposal for an LNG facility in a similar location – the Vista del Sol project. ExxonMobil chose a more formal announcement strategy, staging a press conference, in January 2004 that drew more than two hundred attendees, including Governor Rick Perry, Senator Judith Zaffrini of Laredo, Representative Gene Seaman of Corpus Christi, Texas Railroad Commissioner Victor Carrillo, and Consul General Al-Hayki of Qatar. Governor Perry and other state elected officials attended the initial press conference announcing the project to show their support. "Texas and the United States need secure supplies of natural gas to attract industries, assure development and to continue the strong economic growth we are experiencing in our state and throughout the nation. This project will bring jobs and other economic benefits to San Patricio County and the greater Corpus Christi area and will provide long-term supplies of natural gas for our industries, power plants and homes" (as quoted in Powell 2004).

According to the local community affairs consultant for ExxonMobil, the purpose of this press conference was to inform the community about what ExxonMobil was proposing, build momentum and support for the proposal, and demonstrate broad community support for the proposal to FERC.[9] The company also sponsored three open houses in Ingleside, Port Aransas, and Portland in late March 2004. FERC representatives attended these meetings and reported that, "Most comments received at the open houses were in support of the project. The only concerns expressed were over the breadth of the cumulative impacts analysis and safety" (Federal Energy Regulatory Commission 2004d). Because ExxonMobil followed Cheniere, they were largely able to piggyback off of the efforts of Cheniere in terms of public relations. Many interviewees, including ExxonMobil's community affairs consultant, mentioned the excellent job that Cheniere did in telling the LNG story, particularly to local environmental organizations, in order to secure their support for LNG terminals.[10] A third proposal – OxyChem's Ingleside Energy Center – came on the heels of the Vista del Sol proposal.

All city and state officials were supportive of all three projects during the FERC review. "Unlike other areas of the country, where the prospect of LNG has stirred up controversy about safety, the local community has been supportive. Government, economic development and business leaders and union workers voiced support for the projects at public hearings" (Beshur 2005). Moreover, environmental groups were supportive as well. In comments submitted on the Draft Environmental Impact Statements (EIS) of the Cheniere Corpus Christi proposal, the Coastal Bend Environmental Coalition wrote,

We commend Cheniere for their community exposure. As far as we are aware, there has been no serious opposition to this proposal. ... We do not oppose the proposal covered in the Draft EIS. ... We do not believe the location of this terminal and pipeline in the area will increase terrorist activity. The Port of Corpus Christi is already the 5th largest in the nation and we have the greatest concentration of refineries and petrochemical industries in the nation. The

---

[9] D. McLallen, telephone interview with second author, July 6, 2009.

[10] J. French, interview with second author, May 18, 2009; J. LaRue, interview with second author, May 19, 2009; McLallen interview, July 6, 2009; M. Scott, interview with second author, May 18, 2009; T. Simpson, interview with second author, May 18, 2009.

location of one more energy company should not affect terrorism in our view. It is our understanding that state and local community leaders support this application and we add our support. We request that FERC finalize its review as soon as possible. (Patricia H. Suter as quoted in Federal Energy Regulatory Commission 2004b)

Although there were a few complaints about the Vista del Sol proposal expressed by out-of-town environmentalists and union members, specifically targeting ExxonMobil as a corporation, these were quickly dismissed by the local and state officials.[11] When the mayor of nearby Port Aransas expressed concerns about a potential shoreline erosion problem caused by increased tanker traffic to the facilities, Cheniere immediately wrote a check for $250,000 to match a federal program to prevent erosion. This check came even before the facility was permitted. This action did a great deal to earn trust in the Port Aransas area, the citizens of which could have become the most vocal opponents of the proposal.

In the face of such vocal support for the proposals, it is no surprise that all three were approved relatively quickly by June 2005. However, none of these three facilities have been built. Cheniere began site preparations in early 2006, but the project is currently on hold awaiting more favorable economic conditions. Instead, the company focused its efforts on construction of two other facilities that had already been approved – Freeport LNG and Sabine Pass LNG. Other irons proved hotter for Cheniere. Likewise, ExxonMobil canceled plans to develop its proposal in August 2006 in favor of its Golden Pass project (which was approved just one month after Vista del Sol in July 2005) at Sabine Pass, Texas, which had better connectivity to interstate and intrastate gas pipelines. In early 2007, ExxonMobil sold its permits to develop the Vista del Sol facility to 4Gas, an independent Dutch company dedicated to developing and operating LNG import and regasification terminals. No construction has begun on the project due to unfavorable market conditions, specifically that the Gulf Coast market was already quite saturated with LNG projects by this time. These stories demonstrate the impact that prevailing market conditions and the approval of nearby similar facilities can have on the eventual construction of a given proposal.

---

[11] There was a somewhat significant dispute about the mitigation proposed by ExxonMobil that rallied residents of the nearby wealthy neighborhood of Northshore. However, residents were not against the Vista del Sol proposal but rather its mitigation plan. ExxonMobil quickly withdrew the mitigation plan, and this opposition dissipated.

Mobilization and Rejection

From the point of view of activists, perhaps our most salient finding concerns the relationship between opposition mobilization and project rejection. The first thing to note is that this relationship is not the converse of the link between nonmobilization and approval. That is, although a lack of opposition was more or less sufficient to explain approval, the presence of opposition is not enough – in and of itself – to account for rejection. Our data suggests that mobilization is neither sufficient at a consistency 0.61 nor necessary at a consistency of 0.68 for rejection.

These results beg the question: what other factors are important in driving rejection? Figure 4.1 helps to provide some clues as to these factors, which we then include in an exploratory analysis in the following text. First, of the eight proposals in our sample that were rejected, seven fell into the set of communities that mobilized against the project. The one case that did not follow this trend is KeySpan, which was rejected despite low levels of community mobilization due to an intergovernmental conflict, which we will describe in the following section. This case suggests that intergovernmental conflict over projects could play an important role in eventual rejection.

Second, of the ten cases that fell into the set of mobilization, only seven were rejected. What happened in the other three cases (Aiken County, Cameron Parish Gulf Landing, and Claiborne County)? Aiken County and Claiborne County are the two nuclear-related projects, in which, as described in Chapter 4, mobilization was largely led by groups located outside the locally affected community. Cameron Parish Gulf Landing, which will be described in greater detail in Chapter 5, follows a similar pattern, with opposition coming late in the review process and largely from the environmental and fishing communities in New Orleans – across the state and a world away from Cameron Parish. These groups sued after the project was approved and lost, but the proposing company eventually decided to withdraw the proposal, citing economic considerations. Thus it appears from these results that opposition efforts that are largely driven by groups outside of the affected community are not effective in swaying regulators to reject a given proposal – another factor that we will include in our exploratory analysis of rejection. Instead, local grassroots opposition, particularly when combined with outside forces, appears to be the most plausible route for opposition to lead to eventual rejection. We will see a clear example of this combination in our narrative of the Long Beach case – a case that also included a heated intergovernmental dispute between the California and federal regulators.

TABLE 4.3. *Fuzzy-Set Scores for Conditions in Rejection Analysis*

Case	Intergovernmental Conflict	Outside Leadership (absent local)	Mobilization	Rejection
Aiken	0	1	0.6	0
Brazoria	0	0	0.2	0
Cameron Creole Trail	0	0	0	0
Cameron Gulf Landing	0.8	1	0.6	0
Cameron Sabine Pass	0	0	0	0
Cassia	0	0	0.2	0
Claiborne	0	1	0.8	0
Essex	0.6	0	0.2	0
Gloucester	1	1	0.2	0
Long Beach	1	0	0.6	1
Malibu	0.6	0	0.8	1
Mobile	0.6	0	0.6	0.8
New Castle	1	0	0.6	1
Providence	0.6	0	0.2	1
Riverside	0.8	0	0.6	1
San Patricio Cheniere	0	0	0	0
San Patricio Vista del Sol	0	0	0	0
Ventura	0.6	0	1	1
Whatcom	0	0	0	0

In order to test some of these ideas about other factors, besides mobilization, that may be relevant to rejection, we again turn to fs/QCA. For the purposes of this exploratory analysis, we included all cases except Solano, for a total of nineteen cases (see Table 4.3). No EIS was prepared for the Solano project, and the EIS was an integral component in scoring intergovernmental conflict. In addition to mobilization, we include two other causal conditions related to whether a project gets rejected: intergovernmental conflict and outside leadership of the opposition in the absence of local involvement.

By intergovernmental conflict, we mean conflict between different layers of government, particularly state and federal officials, about technical or legal aspects related to the project. For the most part, relations between state and federal agencies are often neutral when discussing such projects, but sometimes these discussions take on a different tone. For example, in reviewing the Final EIS for the Gulf Landing proposal, the National Marine Fisheries Services (NMFS) offered the following comment:

Given the NMFS role in managing the nation's marine fishery resources, the agency is particularly concerned with [the Coast Guard's and Maritime Administration's] conclusion that the proposed [open-loop] system will have no significant impacts on our red drum and other managed fishery species in the Gulf of Mexico. Furthermore, NMFS remains concerned about the potential for significant cumulative fishery impacts resulting from multiple LNG facilities proposed in the Gulf of Mexico. These possible impacts were not adequately evaluated either in the draft or final EIS. In recognition of the potential significant adverse environmental consequences of authorizing an LNG terminal using an [open-loop] system, the NMFS continues to recommend that [the Coast Guard and Maritime Administration] not authorize such a system, but alternatively require a closed-looped system similar to those presently authorized at onshore and offshore LNG facilities. (National Marine Fisheries Service 2005)

These concerns are often expressed through frequent and lengthy letters about the project during the project's EIS consultation process. In addition, escalating tensions can be referred to the White House Council on Environmental Quality for resolution (as they were in the case of the Gulf Landing project), state agencies may deny permits, and court cases may ensue. We believe that this type of intergovernmental conflict is extremely important for the opposition's success. Not only does it openly display a form of "elite division" that is often seen as important for movement formation in the literature, but it also "certifies" and provides elite allies for (as we will argue in Chapter 5) burgeoning oppositional claims. We measure intergovernmental conflict as follows: a case received a score of 1 if a lawsuit was filed by state or federal officials related to the project, 0.8 if interagency disputes were referred to the White House Council on Environmental Quality and a state or federal agency denied a permit associated with the project,; 0.6 if state or federal agencies wrote letters critical of the proposal during the EIS review process, and 0 otherwise. Thus higher scores on this condition should be associated with rejection.

Outside leadership in the absence of local support also seemed to be an important factor in determining project outcome. Not only can it be dismissed by agency reviewers as irrelevant to local community acceptance, but it can also spur supporters in the local community to speak up and defend the community's right to determine its own fate. For example, in an article in the *Port Gibson Reveille*, editor Emma Crisler described a local meeting organized against the Grand Gulf nuclear proposal as follows, "There were very few local residents at the January 21st meeting and there were many outsiders with no ties to this county who had lots to say about what we should do. I don't know about you, but I resent others telling Claiborne County what we should do, period" (Crisler 2004). To measure this factor, we used our qualitative knowledge of the cases and a reading of EIS comment letters in order to determine how strong outside leadership was in comparison to local opposition. We scored this condition dichotomously: a project received a score of 1 if opposition was led by outsiders absent much local support and a 0 otherwise. Thus our expectation is that lower scores on this condition should be associated with rejection. Table 4.3 provides individual scores for each case; Appendix B provides the relevant information behind each of the scores.

Our results confirm our initial suspicions. Not having outside leadership in the absence of local support proved to be a necessary condition for rejection (1.00). In other words, successful mobilization efforts, in terms of securing project rejection, are often a combination of both external and local opposition. With this necessary condition satisfied, the combination of mobilization and intergovernmental conflict is the best path to rejection, with a consistency of 0.79 and coverage of 0.56. This combination explains six of our seven rejected cases: Mobile, Malibu, Long Beach, New Castle, Riverside, and Ventura. Worth noting, the one case that remains unexplained is KeySpan, which we will describe in more detail in the following text.

Interestingly, if we drop mobilization alone and instead run a condition – *widespread opposition* (i.e., incorporating locals and outsiders) – that combines the opposite of our outside leadership condition and mobilization together (essentially this creates a causal condition that grants a score of 0 to any cases that mobilized through outside leadership alone and grants our original mobilization score to all other cases), we find that the combination of widespread opposition with intergovernmental conflict explains rejection with a consistency of 0.95 and coverage of 0.56. Again, we explain the same six cases but not KeySpan. We now turn to the story of the one proposal in our sample that experienced little community mobilization but was rejected – KeySpan LNG in Providence County, Rhode Island.

Providence County, Rhode Island (KeySpan LNG)

The KeySpan LNG proposal was first announced in October 2003 as an upgrade to an existing LNG facility in Providence, Rhode Island, in order to allow for LNG deliveries by ship. The land for the facility was already owned by KeySpan and housed an existing LNG tank and liquefied petroleum gas tank. In addition, the community immediately surrounding the proposed facility was occupied largely by poorer minority groups. Thus the combination of similar existing industrial development, economic need, lack of community resources, and a federal decision-making process (low political opportunity) due to existing ownership limited community opposition to the proposal.

Local and state officials – including the Providence mayor, Rhode Island governor and U.S. senator – were initially supportive of the proposal. For example, Governor Carcieri testified in front of the Senate Environment and Public Works Committee in March 2004 in favor of the project to address energy shortages and prices, arguing that opposition was a "'knee-jerk reaction' to safety fears that the Coast Guard and other authorities can more than adequately address" (Reynolds 2004). However, this honeymoon period for the proposal was short-lived. As the project progressed through the approval process, nearly every elected official at every level of government came out against the project as being too risky for such a populated area. Interestingly, this change of heart among state-level officials was not driven by local community opposition but instead highly influenced by the strong opposition to a similar proposal – Weaver's Cove LNG – across Narragansett Bay in Fall River, Massachusetts. The Weaver's Cove proposal has attracted some of the most virulent and sustained community opposition of any LNG proposal in the Northeast – opposition that only recently subsided with the company's withdrawal of the proposal in June 2011. The opposition to Weaver's Cove was what first alerted Rhode Island officials to the potential safety issues associated with LNG facilities. In addition, the KeySpan and Weaver's Cove proposals required a similar route for LNG tanker traffic into the bay, passing by many wealthier Rhode Island residents south of Providence.

Most of the opposition to KeySpan was driven by Rhode Island Attorney General Patrick Lynch and his staff, hence, the project's score of 0.6 on intergovernmental conflict. They researched the impacts of the KeySpan proposal, gathered expert-witness testimony, distributed information, and advocated on behalf of the people of Rhode Island – even hiring counterterrorism expert Richard Clarke to conduct a review of the proposal. The attorney general's office did not, however, incite mobilization. Interviews and newspaper articles noted that the response from the community was somewhat subdued compared to past events.[12] Assistant Attorney General Paul Roberti admitted that his strategy was not to garner support from the general public, whom he thought FERC would ignore, but from expert witnesses.[13] Thus the opposition's strategy was to rely on legalistic arguments against the proposal.

After the attorney general became involved in investigating the safety of LNG tankers in Narragansett Bay, nearly all elected officials across party lines turned against the KeySpan and Weaver's Cove projects near the end of 2004 and early 2005. Most officials cited safety concerns about locating LNG facilities in "heavily populated" areas – the same framing used against Weaver's Cove – as the reason for their opposition. Given the proximity to the September 11th attacks and the resultant closure of LNG tanker traffic to the nearby LNG receiving terminal in Boston, opponents were particularly concerned with the possibility of a hijacked plane flying into a tanker, even requesting that such a scenario be included in Clarke's analysis. In addition to state officials, two powerful environmental organizations, Save the Bay and Conservation Law Foundation, also took positions against both projects.

In June 2005, the Weaver's Cove and KeySpan LNG projects were brought before the FERC for a ruling. The FERC determined that, as proposed, the KeySpan facility would not meet current federal safety regulations. Because the existing LNG facility had been operating for thirty years without incident, the design had been exempted from modern safety regulations. However, with the proposed changes, the FERC felt an update was warranted. At the same meeting, the FERC approved the much more controversial Weaver's Cove facility. As a result, more vocal opponents of the KeySpan project turned their attention to the Weaver's Cove facility. This result, in some ways, suggests that the more legalistic strategy selected by Attorney General Lynch was an appropriate one. It also provides evidence of the effect that a similar, nearby proposal – particularly one that is highly contentious – can have on the decision-making process. If we could have found a way to incorporate this cause systematically into our analysis, we would have done so. However, relationships between oppositional groups in nearby communities are not always so cordial, as exemplified by the Essex case. As described in Chapter 3, opponents of the Northeast Gateway facility in Essex County, although located in the same state as Fall River in Massachusetts, did not garner support for their cause from Fall River opponents, as Providence opponents had. Instead, the

[12] R. Gallison, interview with R. Wright, June 23, 2009.
[13] P. Roberti, interview with R. Wright, June 24, 2009.

Essex proposal was offered by Fall River opponents as a potential solution to the safety problems of the Fall River proposal. Although the Essex proposal did generate intergovernmental conflict at the same level of the Providence case, it was approved, suggesting that it is the combination of mobilization and intergovernmental conflict that is most potent in securing project rejection.

Despite the FERC's rejection, KeySpan could have easily proceeded with the proposal and won the FERC's approval by making the required safety modifications. However, the company decided not to "due to the high financial costs, the need to acquire several adjacent properties, and the fact that making the modifications would require taking the existing facility out of service for two to three heating seasons" (Federal Energy Regulatory Commission 2004c, ES-17). This result highlights the importance of intergovernmental conflict in project rejection. However, we also want to show the potency of the combination of mobilization and intergovernmental conflict, so we now turn to the story of the Long Beach proposal.

Long Beach, California (Long Beach LNG Import Project)

The Long Beach LNG Import Project was proposed for the Port of Long Beach, one of the country's busiest ports, in March 2003 by Sound Energy Solutions (SES), a subsidiary of the Mitsubishi Corporation. The city of Long Beach was experiencing a budgetary crisis just prior to the proposal, and in 2002, the city manager was fired after announcing a $90 million shortfall in the city's budget. At the time, there were even concerns that the city would have to declare bankruptcy. As a result, the SES proposal and its potential to increase the city revenue appealed to many leaders. Moreover, city residents had suffered a quadrupling of natural gas prices during California's 2001 energy crisis. The potential for a steady supply of natural gas from the SES proposal was also attractive to city leaders.

According to interviewees, Long Beach Mayor Beverly O'Neill, who served in that position for twelve years, was very much probusiness and supported any projects that would bring jobs to the community. Right before the SES proposal was announced, she had appointed James Hankla to the Harbor Commission. Hankla was a former Long Beach city manager and head of the Alameda Corridor Transportation Authority. The Harbor Commission would serve as one of the lead agencies for the proposal's joint Environmental Impact Report (EIR)/EIS, along with the FERC. Chris Garner, from Long Beach's Gas and Oil Department, felt that this change in leadership at the port was one reason the SES proposal was even considered.[14] Six or seven other companies had previously approached city officials about constructing an LNG facility at the port, and all had been turned away. Interestingly, SES initially approached the Port of Los Angeles with its proposal but was rebuffed. Also, as in the case of several other communities targeted for LNG proposals around the United States, the Port of Long Beach/Port of Los Angeles had also been the site of an attempt to site a LNG facility during the 1970s. However, California's 1977 Natural Gas Act (now defunct) precluded it from consideration due to the proximity of surrounding populations. Although Long Beach had previously showed some ability to mobilize around port issues, the community generally did not have a great deal of capacity to do so. During the late 1990s, Long Beach experienced the loss of two major sources of employment in town – the Navy and McDonnell Douglas, an aerospace firm. As a result, many long-term residents moved out of town. Moreover, many of its residents, especially those who live closest to the port, are "poor and brown. ... To be an activist and speak up at meetings, you have to believe that government works and your voice will be heard. Many Latinos don't."[15] There was a small, core group of active citizens in town. Their last concerted effort had formed the Long Beach Citizens for Utility Reform (LBCUR), in response to the rate increases associated with the California energy crisis. During this effort, they had sued the city over the rate increases. The LBCUR Board had consisted of about fifteen people. However, by the time the SES proposal came along, only four were left. Bry Myown was one of these individuals, and she was convinced by the others to lead the charge against LNG. The other three were largely absent from the opposition effort due to unforeseen personal and health-related issues.

In addition, there was also a larger, regional coalition of environmentalists forming in response to the air-quality concerns related to Port of Los Angeles and Port of Long Beach operations. In 2003, a coalition of residents in San Pedro and Wilmington, aided by the Natural Resources Defense Council (NRDC) and Coalition for Clean Air, had successfully sued the Port of Los Angeles on clean air issues. In September 2004, Port of Long Beach harbor commissioners certified an EIR on a port expansion project (of Pier J) only to have it appealed by the local residents (mostly California Earth Corps), again aided by the NRDC and Coalition for

---

[14] C. Garner, interview with second author, November 3, 2008.
[15] B. Myown, interview with second author, November 4, 2008.

Clean Air, to the Long Beach City Council. Although the city council deferred a decision on the port expansion proposal, the Port of Long Beach eventually withdrew the proposal.

In general, Long Beach residents were becoming increasingly sensitized to air-quality concerns associated with the port and goods movement. A number of scientific reports had been released indicating the existence of a "death zone" surrounding the two ports. This concern actually dovetailed nicely with SES advertisements that LNG could be used to clean the air. As a result, Myown reluctantly chose to frame opposition around safety and terrorism. She didn't feel she could get the Long Beach community to "make the leap" to concerns about climate change and fossil fuel dependence, as had been done by opponents of the nearby Cabrillo Port LNG proposal. Instead, Myown felt the need to "tailor her message" to the Long Beach community.[16]

Unfortunately for Myown, in the case of LNG, NRDC refused to become active in the opposition effort. Myown approached its leadership but was told that the board refused to take a position against the proposal. Myown felt this refusal was associated with its belief that the LNG could possibly be used to convert local vehicle fleets to natural gas and thus potentially improve air quality in the vicinity of the port. In terms of the resources available to the opponents, it consisted of money contributed to the fight by three to four households.

All major decisions about the LNG proposal in the Port of Long Beach would be made by nonelected officials. As mentioned, the lead agencies for the joint EIR/EIS were the Port of Long Beach and FERC, respectively. As the main economic engine in the city of Long Beach, the port provided one in every seven local jobs in the city. As a result, the city council generally tended to stay out of port affairs. However, in the case of EIR certification by a nonelected body like the Port of Long Beach, California law permits an appeal of this certification to the next-highest elected body, which would be the city council. However, the city council cannot disapprove of the project, it can only determine that the EIR is flawed and send it back to the port for revisions.

Because no votes by elected officials would (possibly) be taken on the proposal until after the completion of the EIR/EIS, local opponents chose a legalistic strategy from the very beginning, trying to prevent the completion of the EIR/EIS. Thus their tactics consisted more of contacting decision makers and media outlets directly, writing letters, filing briefs with

[16] Myown interview, November 4, 2008.

FERC and other state agencies, and ensuring legal representation through other environmental groups (Californians for Renewable Energy and California Earth Corps). Although the opponents fought to put the issue on the city council's agenda and succeeded twice, fundamentally they knew the battle would be a legalistic one. For these city council votes and the EIR/EIS public hearings, the opponents mobilized about one hundred people in opposition. They also organized at least two stand-alone protest events, but fewer than fifty people would show up at these events.

This legalistic strategy dovetailed nicely with the efforts of state officials, many of whom were concerned about the potential impacts on public health and commerce that would result from placing a potentially explosive LNG facility in such a busy location.[17] According to the Safety Advisory Report submitted by the State of California about the facility:

Sound Energy Solutions' (SES) proposed LNG import terminal at the Port of Long Beach would be in a densely populated, urban area, and would pose a risk to the health and safety of the approximately 130,000 people living, working or visiting in the area within approximately three miles of the proposed site. There are more than 85,000 people living within three miles of the proposed site, with residential neighborhoods as close as 1.3 miles away in the City of Long Beach and two miles away in the City of Los Angeles. There are also approximately 44,000 people who work within three miles of the proposed site, including approximately 2,000 daily workers at the Port of Long Beach within one mile of the proposed site. Numerous tourist attractions, such as the aquarium, and parks and recreational activities are as close as 1.5 miles away, and downtown Long Beach itself is just two miles away. (California Energy Commission 2005, 6–8)

In fact, SES's Long Beach proposal sparked an important jurisdictional battle between state and federal regulators. The California Public Utilities Commission (CPUC) thought that it should have authority to grant a Certificate of Public Convenience and Necessity for the SES proposal because all the gas would be used within the state of California. The FERC argued that it should have exclusive regulatory authority because of the role that LNG would play in the nation's energy strategy. After SES filed an application only with the FERC, CPUC filed a lawsuit in the Ninth District. However, before the lawsuit was resolved, the U.S. Congress passed the 2005 Energy Policy Act, which granted the FERC exclusive authority to regulate onshore LNG applications. City of Long Beach officials and California's senators (along with senators from other coastal states facing similar proposals) lobbied extensively to prevent the inclusion

---

[17] M. Schwebs, telephone interview with second author, May 19, 2006.

of such language in the Energy Policy Act but lost. They did manage to eliminate language granting the FERC eminent domain authority in such cases. Even if local and state leaders, elected officials, and agency officials supported the SES proposal and LNG importation more generally, they certainly did not want the federal government preempting local control. Thus this issue of federal preemption became a uniting force.

According to Myown, a major turning point in local discussions about the facility occurred when City Attorney Robert Shannon decided to get involved in the federal preemption discussion. Originally, the city had chosen to stay out of the jurisdictional battle because Shannon felt secure that, because the Port of Long Beach owned and would lease the land to SES, local decision makers would hold ultimate authority over the fate of the proposal. From this point of view, if anything, the CPUC lawsuit against the FERC was an attempt to strip the Port of Long Beach of its local regulatory authority. However, Shannon "changed his tune" after Myown and Bill Pearle (editor of the LBReport.com, an independent online newspaper that actively covered the debate) played him a recording of the U.S. Senate Committee on Energy and Natural Resources, Subcommittee on Energy's February 15, 2005 Hearing on LNG. In the hearing, FERC Director of the Office of Energy Projects Mark Robinson clearly requested that the forthcoming Energy Policy Act grant FERC the right to use eminent domain in the siting of LNG facilities. Eminent domain would mean the Port of Long Beach's status as a landowner of the site would not matter. "As gung ho as he [Shannon] had been to stay out of the fray because he perceived the other thing [CPUC's suit against federal preemption of LNG siting] as Sacramento taking away Long Beach's authority over the Port, he then became afraid of the United States taking away his authority over the Port. And that was the first chink."[18]

Thus the jurisdictional battle between federal and state authorities over the siting of SES's proposed facility mobilized state agencies, particularly the CPUC and California Coastal Commission (CCC), and local officials. After the numerous attempts by the CPUC to preserve state jurisdiction were rendered moot by the passage of the Energy Policy Act, both state agencies submitted very critical comments on the Draft EIR/EIS. The CPUC stated that the proposed project should be "rejected," and the CCC, which would have to certify an amendment to the Port of Long Beach master plan for the facility to be built, expressed "serious concerns

[18] Myown interview, November 4, 2008.

about the adequacy of the EIS/EIR, especially with the respect to the document's analysis of public safety and risk. We strongly believe these inadequacies warrant recirculation of the EIS/EIR" (Federal Energy Regulatory Commission 2005). According to one interview with Robert Kanter from the Port of Long Beach, staff members at the CCC did not seem willing to approve any LNG proposal anywhere in the state. The CPUC was explicit in its desire to bring LNG to California; however, it disagreed with the siting in such a densely populated and economically valuable area.[19]

Thus the CPUC became an unlikely ally of local opponents to the proposal in Long Beach. Unlike other members of the Ratepayers for Affordable Clean Energy (RACE) coalition against LNG (to be discussed in more detail in Chapter 5), who had sued the CPUC to prevent actions to ease the flow of LNG into California, Myown was grateful for CPUC involvement and the associated resources and legal expertise that it brought to the review of technical documents and potential future legal battles with the FERC and SES about the Long Beach proposal. Although she was trying to get state legislators to support the CPUC's stance on federal preemption, RACE was trying to get state legislators to support its lawsuit against CPUC. According to Myown, these two simultaneous requests from anti-LNG activists "really muddied the waters" when she would talk with decision makers:

I felt a little stupid during the RACE phone calls because I was bringing up issues the others didn't know anything or care about. . . . The Sierra Club would tell me to put out flyers at the local farmers market. They didn't get my community – Long Beach doesn't even have a farmers market! I also had to become an ally with Michael Peevey [president of the Public Utilities Commission and a strong supporter of LNG generally] even though I didn't agree with most of his policies. I needed the resources of the Commission as a state agency to comb through the Long Beach proposal and EIS. However, that friendship put me at odds with other LNG opponents in the state who were suing them on another issue. In many ways, I felt more of a kinship with opponents facing onshore facilities in the Northeast.[20]

In addition to allying with state officials, Myown also ran for an open council seat in April 2006, in part "to ensure that every candidate, even those in the mayor's race, would be forced to take a position on the proposal."[21] Although she did not win the election, Myown ensured that

---

[19] R. Kanter, interview with second author, November 5, 2008.
[20] Myown interview, November 4, 2008.
[21] Myown interview, November 4, 2008.

the LNG issue was prominent during the campaign with both of the key contenders in mayoral race expressing opposition to the proposal in favor of offshore alternatives. These oppositional stances were bolstered by the fact that SES seemed unwilling to commit to a satisfactory compensation package for the city. City officials expected some sort of large upfront commitment from the company in order to compensate the community for accepting the facility. However, SES never made such a commitment, either in terms of reductions in gas prices for Long Beach consumers, the price of leasing the pipeline capacity from the city, or conversions of vehicle fleets to LNG in and around the port.

Soon after the election of a new mayor who was opposed to the proposal, Port of Long Beach commissioners voted to abandon the EIR process and end negotiations with the proposing company about the LNG facility in January 2007. SES sued to try to force the port to complete its review but lost when a judge dismissed the case in March 2008. Most interviewees seemed to think that the project's demise was related to a combination of community opposition, state agency opposition, and the mayoral election.[22]

CONCLUSION

In Chapter 3 we reported exceedingly modest levels of mobilized opposition across our twenty cases. Despite our hunch that the tendency of scholars to select mature movements for study was exaggerating the amount of contentious collective action in society, even we were surprised by the low levels of activity we found. The results reported in this chapter, however, represent a second surprise. In the face of such modest levels of emergent opposition, we were prepared to find that mobilization had little or no influence on project outcome. We were wrong. Our results support the opposite conclusion. We interpret the results of our analyses of reject and approval as supporting the notion that mobilization, or lack thereof, is among the key causal conditions shaping the ultimate fate of these projects.

Ironically, it is our analysis of the cases that were approved that perhaps affords the most compelling evidence of the link between mobilization and project outcome. As we reported in the preceding text, the absence of a mobilized opposition is more or less sufficient to explain approval. Nine of the ten projects that experienced no mobilized opposition were approved. Bottom line: absent at least some localized opposition, proposed projects –

[22] Garner interview, November 3, 2008; Kanter interview, November 5, 2008; T. Modica, interview with second author, November 3, 2008.

even ones that present serious, objective risks to the environment – are virtually assured of being approved.

Rejection is more complicated and certainly not the simple converse of approval. That is, although moderate to high mobilization is clearly part of the story, it is neither sufficient (at a consistency of 0.61) nor necessary (at a consistency of 0.68) to account for rejection. The following two summary statistics underscore the point.

- Of the eight proposals that were rejected, all but one – Key Span – fell into the set of communities that experienced some level of mobilized opposition.
- Three of the ten communities that "mobilized" had their projects approved.

But if the presence of a mobilized opposition is not enough to explain rejection, the incorporation of one other causal condition into a simple two-factor recipe does the trick. One of the central themes we stress throughout the book is the need to accord other actors far more importance in our theories and empirical accounts of contentious politics. Consistent with that admonition, the combination of mobilization (specifically mobilization that is not only led by those outside the affected community) and intergovernmental conflict represents the best explanatory "path" to rejection. This recipe accounts for six of our seven rejected cases: Mobile, Malibu, Long Beach, New Castle, Riverside, and Ventura. Projects that face mobilized opposition and engender conflict between state actors are extremely likely to be rejected.

Finally, what can we say about "built"? Our conclusions echo those reported for "rejection." Once again, mobilization – or rather its absence – is part but not the whole of the story. The absence of mobilization is not sufficient to explain built, but it comes very close to being necessary for that outcome, with a consistency of 0.75. This finding makes great theoretical and intuitive sense. If, as we reported in the preceding text, nonmobilization is sufficient to explain approval, and approval is, in turn, necessary for a project to get built, then it follows that nonmobilization is also necessary for built. Approval is not at all sufficient to ensure that a project gets built. Beyond approval, there are a number of factors – principally economic in nature – that condition the likelihood of a project being built. By incorporating two important economic conditions into a model with low mobilization, we are able to do a reasonably good job of explaining built. These two factors – designated here as *regional saturation* and *few irons in the fire* – were designed to measure the number of rival LNG facilities active (i.e., already on line or under construction) in the region and other LNG proposals being pushed in the same region by company officials. Taken together, these conditions crudely capture the relative market advantages that attach to each of the projects. High values on both conditions mean that the company may have little incentive to build the project even if it has been approved. When combined with low mobilization, the resulting three-factor recipe explains built with reasonably high measures of consistency (0.81) and coverage (0.75).

Besides the strong and somewhat surprising influence that mobilization (or its absence) appears to exercise in all of our final recipes, the findings reported in this chapter also underscore the need to take other actors and factors into account as we seek to understand the outcome of potentially contentious infrastructure proposals. If the absence of mobilization was sufficient to explain approval, the actions and dispositions of two other critically important categories of actors were implicated in the other two outcome conditions that we studied. Significant disagreements among state actors (e.g., intergovernmental conflict) regarding these projects help explain rejection, while company assessments of the competitive viability of each project (as shaped by regional saturation and irons in the fire) clearly shaped the ultimate decision to build or not to build.

5

From Not in My Backyard to Not in *Anyone's* Backyard

The Emergence of Regional Movements against Liquefied Natural Gas

In the previous two chapters we took up issues – movement emergence and outcomes – to which scholars have devoted a great deal of attention.[1] By contrast, in this chapter, we focus on a topic about which we know comparatively little. We refer to "scale shift," or the geographic expansion or contraction of contention. In a 2005 article, McAdam and colleagues lament what they see as the "distorting lens" of social movement scholarship. More specifically, they fault movement scholars for continuing to propound a stylized view of social movements rooted in the field's early preoccupation with the New Left/New Social Movements of the 1960s and 1970s. This "stylized view," they contend, equates movements with the following four features:

- Disruptive protest in public settings,
- Urban and/or campus-based movement activities,
- Contentious claim making by disadvantaged minorities, and
- Loosely coordinated national struggles over political issues.

They go on to present data from their time-series analysis of "protest events" in Chicago between 1970 and 2000 in an effort to show that all of these features are wildly atypical of movement activity, especially during the most recent period.

In this chapter we focus on the last of these stylized elements. McAdam and colleagues (2005) frame their discussion of the element in the following way: "National movements tend to be the modal object of study for movement analysts. So, for example, for the U.S. we have extensive

[1] A previous version of this chapter was published in Hilary S. Boudet. 2011. "From NIMBY to NIABY: Regional Mobilization against Liquefied Natural Gas in the United States." *Environmental Politics* 20 (6): 786–806.

literatures on such national struggles as: African-American civil rights, women's, environmental, pro-choice, pro-life, among others. Does the empirical attention accorded [these] national movements match their actual proportion in the population of all contentious events?" (10). Based on their data, the answer is unambiguous. "A scant six percent of all ... events were coded as 'national' in their focus .... In contrast ... nearly three-quarters of the total were judged to be focused on 'city' ... or neighborhood issues" (10–11).

The vast majority of movements and preponderance of movement activity then would appear to be local in its geographic/jurisdictional focus. But it isn't simply that national struggles are overrepresented in empirical studies of social movements. In addition, even the concept of national movement rests on something of a fiction. With rare exceptions, what came to be seen as national movements began life as local struggles. Although the Montgomery Bus Boycott is properly seen as catalyzing the mass movement phase of the civil rights struggle, it is clear that the participants initially saw themselves as involved in a localized episode of contention. Yet we know very little about the process by which Montgomery was transformed from a local dispute into the cutting edge of an emerging national movement.

In short, movement scholars have exaggerated the frequency of national movements and glossed a critically important process in the development of true national struggles. We at least have a name for this process. Tarrow and McAdam (2005) use the term *scale shift* to describe the process by which movements grow beyond their localized beginnings. Despite this conceptual work, however, there have been few systematic studies of the phenomenon. Why do most movements remain narrowly local in focus? How do conflicts initially defined as local come to be seen as relevant to groups elsewhere? And what determines the scale to which movements shift? Or put another way, once the process of upward scale shift starts, why don't all movements scale up to the national level? With this last question we mean to call into question the overly simplistic distinction between local and national movements. There are many more levels of movements than just these. In the United States, we can speak meaningfully of at least six levels of movements: neighborhood, town- or citywide, state, regional, national, and transnational. The point is that "local" movements come in various shapes and sizes. Montgomery began as a citywide movement. The 1969 Stonewall Riot that gave birth to the gay liberation movement was a neighborhood altercation between gay bar patrons and the New York City police. The lunch counter sit-in in Greensboro, North Carolina, that spawned the 1960 regional sit-in movement was initially a strictly local neighborhood event. Although all three of these cases eventually scaled up to the national level, there is nothing inevitable about that process.

Reflecting widespread concern with the technology during the period of our study, thirteen of our twenty cases involved efforts to block the construction of liquefied natural gas (LNG) terminals at various sites all over the country. But despite the fact that the issue was national in scope, we find no evidence that these thirteen cases (or the several dozen others that were unfolding at the same time) were ever linked into anything that resembled a national anti-LNG movement. At the same time, the cases also do not appear to be thirteen discrete local struggles. The reality looks to be something in between these two extremes. The majority of the cases can be grouped into two distinctive regional movements that developed in opposition to LNG along the Gulf Coast and West Coast, respectively. No such regional movement, however, emerged in the northeastern United States. Despite occasional efforts by activists to link to struggles in other locales, in the end the Northeast cases remained discrete local conflicts. Taking advantage of this variation, we use the comparative analysis reported in this chapter to shed systematic light on what we see as the dynamic mechanisms that shape and constrain the spread of contention.

THE SPREAD OF CONTENTION: WHAT DO WE KNOW?

Despite the proliferation of scholarship on social movements/contentious politics throughout past thirty years, there are still a number of glaring holes in the literature. One of the more significant gaps in knowledge concerns the factors and dynamic processes that shape the geographic expansion (or occasionally, contraction) of a movement. The dirty little secret is that our theories of movement emergence rest on a simplifying fiction: that movements are born fully formed. So to revisit the examples touched on in the preceding text, we are asked to believe that the civil rights movement emerged in Montgomery during 1955 and 1956 or that gay liberation was "born" the night of the Stonewall riot, and so on. The more complicated truth is that these local struggles only became national movements as they inspired or linked up with similar efforts elsewhere. The problem is we know very little about this transformative process. We can only think of three very small and specialized literatures that bear on the issue.

From Not in My Backyard to Not in Anyone's Backyard

Perhaps the closest literature to what we study here relates to research on Not in My Backyard (NIMBY) movements against locally unwanted land uses like prisons, landfills, power plants, and mental health facilities. However, work on NIMBY movements has mainly focused on explaining local opposition. Although some scholars have gestured to the existence of Not in Anyone's Backyard (NIABY) movements (see, e.g., Lesbirel 1998), few have tried to explain how this upward scale shift occurs. Walsh, Warland, and Smith (1997) provide some clues on the important factors for scaling up in their analysis of eight antiincinerator movements in the Northeast. In seeking to explain how local opposition spreads to the county (against a single project), the authors stress the significance of one mechanism in particular. They term the mechanism *frame expansion*, but it is really identical to what Snow and Benford (1992) have called *frame bridging*. Frame bridging/expansion takes place when a particular collective action frame is successfully applied to a seemingly separate issue or conflict. In most cases of successful scale shift studied by Walsh and colleagues, the antiincineration frame was combined with another frame (e.g., an area's negative image as a wasteland) in order to appeal to a broader constituency. Given the prominence accorded the mechanism by Walsh and colleagues, we also pay close attention to frame bridging/expansion in our analysis of the LNG cases.

One reason why this phenomenon of shifting from NIMBY to NIABY may not be discussed much in the literature is that it is rare. The truth of the matter is that it is hard enough to mobilize people against things going on in their own backyard, let alone what is going on in a neighboring town or across a broad region. In addition, as Piller (1991) notes in his comparative work on NIMBY movements in the United States, regional environmental groups would seem to be "natural allies" for local NIMBY movements and thus vehicles for scaling up, but mutual suspicion has often undermined this potential. As Piller (1991: 166) writes,

NIMBYs and environmentalists would seem like natural allies. In practice, they have had an overlapping yet uneasy relationship. Environmentalists tend toward a big-picture critique of society, or at least hold a vision of sweeping policy reforms designed to preserve the natural environment. In contrast, environmentally-based NIMBYism almost always stems from personal experience of technology's intrusions on everyday life .... NIMBYs don't share the counterculture beliefs of environmentalists. In fact, many NIMBYs dislike them .... NIMBYs also have a keen sense of urgency that separates them from mainstream environmentalists ....

NIMBYs distrust experts as too likely to compromise. Mainstream environmentalists speak the language of dominant institutions and are focused on gradual change, [while] NIMBYs look to disrupt official proceedings.

Thus we have few examples of national antiinfrastructure movements in the United States; antinuclear and antiincineration being the notable exceptions (Rootes 2009a). Interestingly, when such movements have managed to develop, they have been quite effective. No new nuclear plants have been constructed in the United States in some thirty-five years. Still, as we note, these are the rare exceptions. Accordingly, the literature on NIMBY movements provides little information about what factors are important for scaling up from local to regional.

Movement Diffusion

The single "largest" body of work on the spread of contention is the literature on the diffusion of movement tactics. But even these studies are not typically framed as work on the expansion of a movement so much as the adoption and spread of a particular "tactical innovation" within or between movements. Some examples will help make the point. Several scholars have sought to study the spread of the sit-in tactic by civil rights activists during the spring of 1960 (Andrews and Briggs 2006; McAdam 1999 [1982]; Morris 1981; Oberschall 1989). But this example can be conceptualized as either the diffusion of a new tactic within an established movement (e.g., the civil rights movement) or the transformation of a single local demonstration into a regional movement (e.g., the sit-in movement). Given the moribund state of the civil rights struggle on the eve of the first sit-in, we think it more accurate to think of the case as a new movement that revitalized the broader civil rights struggle. But this is definitely not how it has been viewed by most scholars.

In a similar vein, Sarah Soule has studied the diffusion of the "shanty-town" protest in the student antiapartheid movement of the 1980s (1995, 1997). But again, situating the work in the diffusion literature, she conceptualized her research as a study of the adoption of a particular innovative new tactic by movement activists, rather than the geographic expansion of a movement per se. Scholars have also studied the diffusion of other tactics or forms of contention including riots (Bohstedt and Williams 1987), strikes (Conell and Cohn 1995), and collective violence (Myers 2000; Pitcher, Hamblin, and Miller 1977). Even recruitment through preexisting network ties has been framed as another instance of movement diffusion, with activism seen as the innovation that spreads as a result of individual or organizational ties (Gould 1991; McAdam 1986).

Less common are studies of diffusion between movements. McAdam and Rucht (1994), for example, framed their analysis of the influence of the American New Left on the rise of the German student movement of the 1960s as an instance of the "cross national diffusion of movement ideas." More generally, McAdam (1995) used the diffusion literature to argue for a new perspective that differentiated between "initiator" and "spin off" movements. Through standard diffusion processes, the former help to inspire or shape the latter.

Although useful, the work on movement diffusion suffers from two problems. First, as previously noted, the work is rarely conceptualized as research on the geographic expansion of a movement. Second, besides noting the structural conditions (e.g., preexisting social ties) that facilitate diffusion, there is otherwise very little attention paid to the social processes that might play a role in movement expansion. To our knowledge, other than the traditional and now discredited work on contagion (Blumer 1939; LeBon 1897; Tarde 1903), the only explicit attempt to theorize the geographic expansion of a movement is Tarrow and McAdam's (2005) work on scale shift.

Scale Shift

Drawing on their earlier work with Charles Tilly, Tarrow and McAdam define *scale shift* as "a change in the number and level of coordinated contentious actions leading to broader contention involving a wider range of actors and bridging claims and identities" (2001: 331). In their 2005 article, Tarrow and McAdam distinguish between two distinct causal "pathways" that can lead to scale shift and identify a number of mechanisms that they see implicated in each pathway. Reflecting on our LNG cases, we attach special significance to two of the mechanisms identified by Tarrow and McAdam.

Brokerage

As used by McAdam, Tarrow, and Tilly (2001), *brokerage* refers to the conscious, strategic effort by movement actors to establish links between two or more previously unconnected social sites. In the case of scale shift, the goal of brokerage is to connect and encourage new actors to affiliate and begin acting in concert with an expanding movement.

Relational Diffusion

By *relational diffusion* Tarrow and McAdam mean "the transfer of information [and coordinated action] along established lines of interaction" (2005: 127). This is the key mechanism implicated in most of the classic studies of the diffusion of innovations. The likelihood and speed of the adoption of the innovation has been consistently shown to be mediated by the strength and extensiveness of ties linking innovators to potential adopters.

We wish to add one final mechanism to these two – or rather three including the aforementioned frame bridging – giving us a total of four around which we will structure our narrative analysis of the LNG cases. The last of our mechanisms is certification.

Certification

Yet another mechanism defined and deployed by McAdam, Tarrow, and Tilly (2001) in *Dynamics of Contention* is certification. According to the authors, *certification* "entails the validation of [contentious] actors, their performances, and their claims by external authorities" (121). Although not specifically used by McAdam and colleagues to explain the expansion of contention, our research has convinced us that the mechanism played a critically important role in the development of regional movements against LNG on the Gulf Coast and West Coast.

A WORD ABOUT OUR CASES

Having provided, in Chapter 2, a detailed description of the procedures we used to select our full sample of twenty cases and descriptive information on each case, we are spared the need to revisit that material here. Instead, we will simply remind the reader about which of our cases involve LNG projects and then briefly discuss the broader population of such projects that will figure in our narrative analyses of opposition to LNG in the Gulf Coast, West Coast, and northeastern United States. We begin by briefly revisiting that subset of our cases that focused on LNG projects.

We examined thirteen LNG proposals from across the country. With the exception of the three California cases,[2] all the LNG proposals

---

[2] The Mare Island Energy and Cabrillo Port projects were studied initially as in-depth cases in order to understand the processes at work in LNG proposals better. The Long Beach proposal was included as another easily accessible, California-based case to test new procedures for data collection and fieldwork.

TABLE 5.1. *Selected Cases*

Proposal	Proponent	Location		Projected Cost	Proposal Timeframe	
		Community	Onshore or Offshore		Start	End
Freeport LNG	Freeport LNG Development	Freeport, TX (Quintana Island)	Onshore	$300 million	September 2001	June 2004
Mare Island Energy	Bechtel and Shell	Mare Island (Vallejo, CA)	Onshore	$1.5 billion	May 2002	February 2003
Long Beach LNG Import	Sound Energy Solutions	Long Beach, CA	Onshore	$350 million	March 2003	June 2008
Sabine Pass	Cheniere Energy	Cameron Parish, LA	Onshore	$500 million	May 2003	December 2004
Corpus Christi LNG	Cheniere Energy	Corpus Christi, TX (San Patricio County)	Onshore	$500 million	May 2003	April 2005
Cabrillo Port	BHP Billiton	Malibu/Oxnard, CA	Offshore (14 miles)	$550 million	August 2003	May 2007
Gulf Landing	Shell	Cameron Parish, LA	Offshore (38 miles)	$700 million	October 2003	March 2007
Vista del Sol	ExxonMobil	Corpus Christi, TX (San Patricio County)	Onshore	$600 million	October 2003	June 2005
KeySpan	KeySpan	Providence, RI	Onshore	$100 million	October 2003	June 2005
Crown Landing	BP	Wilmington, DE	Onshore	$500 million	December 2003?	March 2008
Compass Port	ConocoPhillips	Mobile, AL	Offshore (11 miles)	$500–$800 million	March 2004	June 2006
Northeast Gateway	Excelerate Energy	Gloucester, MA	Offshore (13 miles)	$200 million	June 2004	May 2007
Creole Trail	Cheniere Energy	Cameron Parish, LA	Onshore	$900 million	January 2005	May 2006

included in this analysis were randomly selected in the context of the larger research project described in Chapter 2. As detailed in that chapter, our case selection procedures resulted in the identification of forty-nine proposals that constituted the population from which projects would be drawn. We then randomly sampled eighteen projects for inclusion in the larger research project on energy facility siting. Ten of these eighteen projects were LNG proposals. For the analysis presented in this chapter, we include these ten LNG proposals and the three California proposals (Mare Island Energy Project, Cabrillo Port, and Long Beach). Information about each of these cases is provided in Table 5.1. The start date in the table corresponds to either the first newspaper article on the project or the official filing of the project with the relevant federal agency, whichever came first. The end date in the table corresponds to the date of the proposal's rejection, withdrawal, or approval. Table 5.1 shows that our sample of LNG proposals provided us with a mix of projects from three regions: the Gulf Coast, West Coast, and Northeast. Therefore, we focus our analysis on these three regions.

The reason that our randomly selected sample included so many LNG cases is that there was something of a feeding frenzy on the part of the oil companies to site such facilities during the early 2000s. Due to increasing demand and dwindling domestic supplies of natural gas, as well as a helping hand from the George W. Bush administration, more than fifty such facilities were proposed in U.S. coastal locations between 2000 and 2005. In addition to our thirteen in-depth cases, we include information about many of these facilities in the analysis of regional LNG movements presented in this chapter. To supplement our in-depth case analysis, we used newspaper reports, Environmental Impact Statement (EIS) documentation, Internet searches, and sixteen additional interviews (ranging in time from 30 minutes to 2 hours with an average length of 54 minutes) with anti-LNG activists and agency officials. These additional interviews were intended to give us a broader perspective on the controversy over LNG as it developed within the United States during this period. Including these "supplementary" sessions, we conducted a total of 123 interviews as part of the fieldwork portion of the research.

REGIONAL OPPOSITION TO LNG

From localized beginnings, distinctive regional anti-LNG movements developed in the Gulf Coast and West Coast regions while Northeast region opposition remained highly site specific. To show how regional anti-LNG movements grew in the Gulf Coast and West Coast but not in the Northeast, we provide a detailed narrative of events in each region, with special attention to the key mechanisms of frame bridging, relational diffusion, brokerage, and certification. In the case of the Northeast, we also discuss two specific contextual factors that appear to have also impeded the development of a more coordinated movement against LNG in the region.

Gulf Coast

The Gulf Coast was the first area of the country to experience proposals for onshore and offshore LNG terminals. This is no surprise given the Gulf Coast's long history of accepting and encouraging oil and gas development on and near its shores (Freudenburg and Gramling 1993, 1994; Gramling 1995; Gramling and Freudenburg 1996, 2006).[3] Given this history, it is probably also not surprising that, beginning in 2001, the earliest proposals for onshore LNG facilities in the Gulf Coast occasioned little or no opposition. In fact, eleven of the thirteen proposals for onshore LNG facilities in the Gulf Coast were accepted with open arms.[4]

Accordingly no one probably foresaw opposition of any kind, let alone a concerted regional movement against the "open-loop," offshore LNG proposals announced between 2002 and 2005. Our analysis will focus on the nine offshore facilities proposed for the Gulf of Mexico during this period (see Figure 5.1). The term *open loop* refers to the system of regasifying the LNG. Open-loop systems use warmth from seawater to regasify LNG, while closed-loop systems use either a portion of the imported gas or ambient air. The use of open-loop regasification, which was only proposed for U.S. LNG projects in the Gulf of Mexico, became a major source of contention in the Gulf Coast due to the potentially devastating impacts to Gulf fisheries from entrainment and impingement of fish eggs and larvae in the water intake and the discharge of significantly cooler water after rewarming the LNG.[5]

The first offshore LNG facility proposed in the Gulf of Mexico was the Port Pelican project in late November 2002. This proposal was quickly

---

[3] Florida is a noted exception to this statement.

[4] Noted exception: Mobile, Alabama, and to some extent Quintana Island, Texas.

[5] Some onshore facilities initially proposed open-loop regasification systems. However, given the sensitivity of inshore estuarine environments and warnings from regulators, companies quickly switched these proposals to other forms of regasification that did not use seawater. Data about fish, egg, and larvae populations in offshore areas is much less certain (G. Crozier, interview with second author, April 10, 2009).

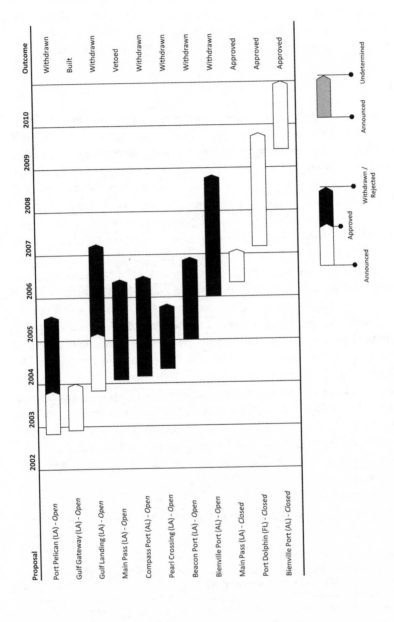

FIGURE 5.1. Offshore LNG proposals in the Gulf of Mexico

followed by the Gulf Gateway project in December 2002. Both projects were to be located offshore in Southwest Louisiana and would use an open-loop regasification system. These two initial projects sailed through the Maritime Administration's approval process: Port Pelican was approved in less than a year and Gulf Gateway in a little more than a year. In these cases, the sole detractors were the National Marine Fisheries Services (NMFS) and a group of ten environmental organizations based out of Southern California.[6] Both wrote letters expressing concerns about the open-loop system, but there was no broader reaction from Gulf Coast residents.

When four more proposals for offshore LNG facilities in the Gulf were announced in late 2003 and early 2004, NMFS personnel became increasingly concerned. NMFS Assistant Regional Administrator of the Southeast Regional Office, Miles Croom, explained, "By the second or third proposal, not only were we becoming worried about the impacts of a single facility but also what might be the cumulative and synergistic impacts of having many of these types of facilities operating in the Gulf at the same time."[7] Regional and state fisheries regulatory bodies – for example, the Gulf of Mexico Fisheries Management Council, Gulf States Marine Fisheries Commission, and Louisiana Department of Wildlife and Fisheries – also became apprehensive about open-loop systems.

These agency officials began to present their concerns at annual public meetings, in which fisheries and environmental organizations were present, and to encourage these groups to become involved. For example, Jeff Rester of the Gulf State Marine Fisheries Commission, remembered personally alerting Cynthia Sartou, Director of the Gulf Restoration Network, to the risks posed by open-loop regasification at the commission's October 2004 meeting and recommending that her organization look into it.[8] The Gulf Restoration Network eventually became an active opponent of open-loop LNG throughout the Gulf Coast. Fisheries regulators thus used existing channels of communication (i.e., annual meetings with stakeholders) – or relational diffusion – to broker an alliance against open-loop LNG involving two previously unconnected groups (fisheries and environmental organizations).

---

[6] The Southern California groups wrote a letter in opposition to the Port Pelican proposal because they were fighting an offshore, open-loop proposal by the same company, ChevronTexaco, just across the California border in Mexico. According to informants, there was no attempt at the time to connect with Gulf Coast groups on the issue.

[7] M. Croom, telephone interview with the second author, April 15, 2009.

[8] J. Rester, interview with second author, April 8, 2009.

The Gulf Landing Final EIS, released in early December 2004, was the first document to include a standardized methodology to assess potential fisheries impacts from open-loop regasification. The range of potential impacts to fish populations was large, and the upper limit[9] caught the public's attention. As a result, in addition to eliciting the same negative comments from fisheries regulators, the Gulf Landing Final EIS also drew negative comments from the Sierra Club's Louisiana Delta Chapter and the Gulf Restoration Network. Fishing organizations were also becoming concerned as local fishermen began to swamp message boards on a popular fishing site, RodNReel.com, and inundate regulators with calls about the issue. The Louisiana Charter Boats Association decided to take on the cause. Environmental and fishing groups eventually joined forces to form the Gumbo Alliance against open-loop LNG in Louisiana after two leaders, Aaron Viles (of the Gulf Restoration Network) and Charlie Smith (of the Louisiana Charter Boats Association) appeared on a local radio show, *Outdoors with Don Dubuc*, to discuss the issue. Despite this opposition, the Gulf Landing facility was approved in February 2005, and Louisiana Governor Kathleen Blanco elected not to use her veto power against the proposal.

This setback did not deter the growing opposition movement. Instead, the Sierra Club, Gulf Restoration Network, and Louisiana Charter Boats Association, with the help of the Tulane Environmental Center, sued the federal government for approving the Gulf Landing facility.[10] They also organized a "boat parade" around the Shell Building in New Orleans and flew a banner over the Shell-sponsored New Orleans JazzFest that read, "Thanks for the music – Don't kill our fish." The escalating movement and growing opposition to open loop did not go unnoticed by elected officials in the region. This time, when the Freeport McMoRan's Main Pass LNG proposal came up for approval in May 2006, Governor Blanco chose to use her veto, citing concerns about open-loop regasification.[11] Blanco's action gave the movement its first real victory and, as such, marked a

---

[9] As an example of the magnitude of the impact, under "worst-case conditions," the Gulf Landing Final EIS estimated a loss of 1.4 million pounds of redfish per year – the equivalent of 8% of annual redfish landings, or catch brought ashore, for the entire Gulf of Mexico. This estimate was later drastically decreased, due to a computer error, to 3.8% of landings lost. However, the damage in terms of public opinion had already been done (National Marine Fisheries Service 2005; Schleifstein 2005).

[10] The lawsuit against Gulf Landing failed, but Shell eventually withdrew the approved facility in March 2007.

[11] Freeport McMoRan came back a few days later with a proposal for a closed-loop regasification system that was eventually approved.

decisive turning point in the conflict. In vetoing the proposal, Blanco served to decisively and publicly certify the movement, granting it a rare kind of legitimacy in a region that was long friendly and beholden to big gas and oil.

Besides energizing the movement, Blanco's certifying veto also set an important precedent for other elected officials in the region. In June 2006, the governors of Alabama and Mississippi, facing similar pressure from environmental and fishing organizations in their own states, joined ranks with Blanco and sent analogous letters to the Maritime Administration promising to veto any open-loop LNG proposals that reached their desks. As a result, ConocoPhillips withdrew its Compass Port proposal just days before Alabama Governor Bob Riley's planned veto. All subsequent open-loop proposals in the Gulf of Mexico have been withdrawn.

Mobilization against open-loop LNG in the Gulf Coast provides a clear example of a regional movement. Opposition went from nonexistent against the two initial offshore, open-loop proposals, to a lawsuit and boat parade against the approved Gulf Landing facility, to a gathering of four hundred individuals at a meeting against Compass Port designed to provide Governor Riley with the public support he needed to veto the facility. In addition, although the Gulf Landing opposition developed after the project's approval, opposition against Main Pass and Compass Port was established well before. Thus mobilization against subsequent LNG proposals in the Gulf Coast became quicker and more intense. As discussed in the following text, brokerage by fisheries regulators and, to a lesser extent, the media created the "unlikely alliance" of environmentalists, recreational fishermen, and commercial fishermen that ultimately carried the day. Key to the alliance was the creative framing of the issue by these brokers that linked the disparate concerns of environmentalists and fisherman. At the same time, relational diffusion proved important for growing the movement from local to regional chapters of existing organizations and sharing tactics and lessons learned across projects. Finally, as noted in the preceding text, the certification of the issue by elected governors in the region granted legitimacy to the issue and undermined the longtime impunity and influence of the gas and oil companies. We amplify these conclusions in the balance of this section.

Brokers played a critical role structurally and ideationally. Structurally, they aggressively brokered ties between two groups – environmentalists and fisherman – who had long been not simply disconnected but openly hostile toward each other. Ideationally, the brokers forged the alliance by creatively joining the concerns of the two groups into a single overarching frame. The framing of the potential threat to the Gulf Coast fishery posed by open-loop regasification technology was used across multiple cases and had implications beyond the locally affected communities. Fisheries regulators provided this initial frame and served as brokers (with the help of the media) to bring together the oft-at-odds fishing and environmental communities against a common enemy. Opponents in different cases focused their framing on open-loop regasification with slogans like "Close the loop" and "Don't kill our fish." For residents throughout the Gulf Coast, the impact on the fishery was a salient issue, as was the threat to sport fishing. As we were told repeatedly in multiple interviews, fishing is a beloved pastime there with a seeming majority of residents owning a boat or "fish camp" where they relax on weekends.[12]

In our view, the actions of the fisheries regulators were the key to the development of the movement. We have already noted the important broker and frame bridging function played by the regulators. But it is important to note that they too afforded the movement a kind of certification at a much earlier point in time than Governor Blanco's 2006 veto of the Main Pass LNG proposal. As "neutral" experts, the regulators enjoyed a form of public standing that the environmental activists could never hope to command. Had the environmentalists reached out directly to the fisherman, they would almost certainly have been rebuffed. The standing of the regulators certified the open-loop issue, creating the possibility for coalition where none had previously existed. The fisheries officials didn't simply create the possibility of coalition but actively sought to forge alliances between established environmental and fishing groups. These alliances proved important because fisheries regulators were able to slow down approval processes, allowing the opposition movement time to grow and provide necessary technical information about potential impacts to the fishery to opponents. Initially, the Maritime Administration adhered to the strict 365-day timeline specified in the Marine Transportation Security Act. However, during the review of the Gulf Landing facility, federal officials, facing pressure from fisheries regulators, determined that they could "stop the clock" to conduct more thorough analyses. Steve Heath, Chief Marine Biologist (now retired) of the Alabama Department of Conservation explained,

---

[12] E. Lamberth, interview with second author, April 9, 2009; L. March, telephone interview with second author, December 10, 2008; G. Russell, interview with second author, April 9, 2009; A. Viles, interview with second author, January 16, 2009.

Reviewing the first two LNG facilities felt like trying to catch a galloping horse. They happened without people even realizing what was going on. By the third one, we had learned to try to loop the horse before it got out of the barn .... I met with some of the state regulators from Louisiana at the Gulf States Marine Fisheries Commission biannual meeting in 2004 .... We went through elected officials in Congress who talked to the Coast Guard about finding a way to slow the process down. Someone finally figured out at the federal level that the regulatory clock could stop until requirements were met. Before that, the Coast Guard wouldn't stop the process. They just talked in terms of the year timeline.[13]

Fisheries regulators also provided an important source of technical information to opponents. For example, when the Gulf States Marine Fisheries Commission drafted a harshly worded letter about the Gulf Landing Final EIS, Rester, an employee, "made the mistake of coming into the office one day over the [2004] Christmas break. My boss asked me to sign and send off our letter on the [Gulf Landing] Final EIS. When I came back in January, I was talking quite a bit to fishermen and other interested parties .... For at least two weeks straight it was one to two calls a day on the [open-loop] issue."[14] Thus, fishermen could rely on regulators to provide them with information about the potential impacts of open-loop LNG.

Although brokerage by fisheries regulators and the media played an important role in frame bridging and coalition building, relational diffusion proved vital for growing opposition from local to regional chapters of existing organizations and sharing tactics and lessons learned across projects. In terms of growing opposition from local to regional chapters, Charlie Smith, the statewide lobbyist for the Louisiana Charter Boats Association, was first alerted to the open-loop regasification issue by two charter boat operators in Cameron Parish. Even the generally conservative Coastal Conservation Association, a national organization of recreational fishermen, lent its voice to the opposition of open-loop proposals after initial hesitation (due to its close ties to the Bush administration) because of the increasing pressure the association felt from its Gulf Coast membership. The Coastal Conservation Association lobbied against open-loop LNG in the Gulf at the state and national level.[15]

Part of the reason for the slow initial response to the LNG proposals on the part of the environmental community is that early proposals were located offshore Southwest Louisiana, where the environmental

[13] S. Heath, interview with second author, April 10, 2009.
[14] Rester interview, April 8, 2009.
[15] Lamberth interview, April 9, 2009.

community has weak ties and limited local leadership, thus hampering efforts at relational diffusion. Leslie March of the Sierra Club's Louisiana Delta Chapter explained,

The Sierra Club was totally asleep on the permits for the initial proposals. We have a small number of people working on so many issues, and every region of the state has severe environmental and health issues . . . . The Acadian Group in Southwest Louisiana [a chapter of the Sierra Club] should have picked up on the issue, but they mostly focus on the Atchafalaya Basin and, although the Sierra Club has members in Lake Charles, there is no real leadership.[16]

It was not until after the Main Pass proposal, located near New Orleans, was announced in February 2004 that many of Louisiana's New Orleans–based environmental groups became more heavily involved in the issue. This highlights the role that strong local chapters play in alerting their regional counterparts to scalable issues.

In terms of sharing tactics and lessons learned, established regional environmental groups relied on existing channels of communication. For example, the Gulf Restoration Network and Sierra Club surfaced in opposition efforts against several proposals. To link opposition activities across multiple states, the Gulf Restoration Network received a grant from the Pew Charitable Trusts that enabled them to hire six Green Corps volunteers to work against open-loop LNG projects. In addition, Mobile Baykeeper Executive Director Casi Callaway also served on the board of the Gulf Restoration Network and could easily apply lessons from the experiences of opponents in Louisiana to the fight against Compass Port in Alabama. For example, utilizing tactics that had proven successful in Louisiana, Callaway brought together an "unlikely alliance" of commercial fishermen, recreational fishermen, and environmentalists (the Gumbo Alliance in Louisiana was reborn as the Gulf Fisheries Alliance in Alabama). The improbability of such an alliance was demonstrated by the fact that the commercial and recreational fishing groups so hated working together that their representatives requested to be placed on opposite sides of the podium during joint press conferences. Moreover, focusing political pressure on the governor, a tactic that had proven successful in Louisiana, was again used in Alabama through constant letter writing, faxing, e-mailing, and lobbying by fishermen.

The mobilization effort against open-loop LNG projects in the Gulf Coast thus provides a clear example of a regional movement. Our analysis

---

[16] March interview, December 10, 2008.

shows that fisheries regulators and the media played a key role in providing a salient bridging frame for the issue (in terms of the impact to Gulf Coast fisheries) and brokering an unlikely alliance (among environmentalists, commercial fishermen, and recreational fishermen). The environmental community, specifically the Gulf Restoration Network, then picked up the issue and through relational diffusion, engaged in similar mobilization efforts against subsequent open-loop LNG projects throughout the Gulf Coast. Once established, the movement, with help from fisheries regulators and the veto power afforded adjacent state governors, prevented the approval of additional open-loop LNG proposals in the Gulf of Mexico.

West Coast

On the West Coast, the anti-LNG movement grew out of localized opposition to onshore facilities and eventually became a regional effort against LNG generally.[17] Twenty LNG projects were proposed from Baja California to Vancouver, Canada, between 2002 and 2009 (see Figure 5.2). The first proposals were announced in early 2002, and by mid-2003 energy companies had submitted seven onshore proposals in California and Mexico. Unlike in the Gulf Coast, every one of these proposals experienced quick, localized opposition movements. For example, throughout the course of nine months in Vallejo, opponents of the Mare Island Energy Project generated almost two hundred letters, raised $15,000, opened a storefront downtown, and organized four different protest events, including an anti-LNG poetry reading (Boudet 2010). Unlike in the Gulf Coast, the main framing of threat used by local opponents against these initial onshore proposals focused on the potential safety and terrorism risks posed by the facilities, as well as local environmental and land-use concerns, arguing that LNG should be moved offshore.

The activities of local opposition groups began to catch the attention of energy companies (that started to propose offshore projects in mid-2003) and regional environmental organizations. In Northern California, the San Francisco–based environmental social movement organization Pacific Environment brokered connections between opponents of Shell's Mare Island LNG Project in Vallejo and opponents of Shell's Sakhalin-II project

---

[17] Carruthers (2007) has written about the anti-LNG movement in Baja California. For this reason, we focus most of our analysis on efforts in the United States.

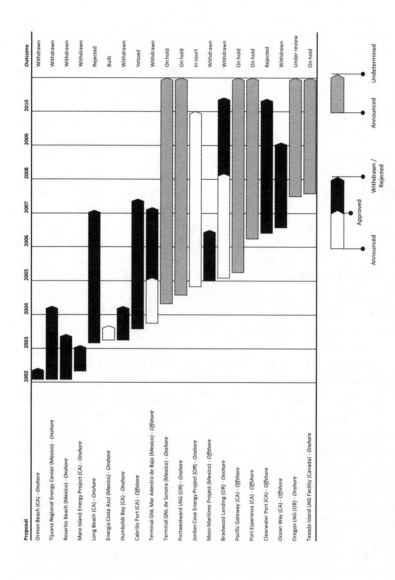

FIGURE 5.2. LNG proposals on the West Coast

in Russia, one of the likely sources of the LNG. According to Rory Cox of Pacific Environment, he viewed the Vallejo project as an opportunity for his organization "to educate the [San Francisco] Bay Area about what was going on in Russia."[18] Thus Pacific Environment began to plant the seeds of a larger framing of the threat posed by an LNG facility – one that moved beyond local safety concerns to opposition to LNG, more generally due to larger supply chain considerations. In Southern California, the Border Power Plant Working Group, an organization started by Bill Powers and Carla Garcia-Zendejas to "accelerate the development of renewable energy resources in the US/Mexico border region," began to help local communities fight LNG proposals in Baja, bringing technical expertise and larger frames about the need (or lack thereof) for LNG (Border Power Plants Working Group 2009).[19]

Still, opposition remained overwhelming local and isolated. Northern and southern opposition efforts on the West Coast were not linked until December 2003 when Cox received a phone call from Powers. Powers had spoken with Loretta Lynch, then a California Public Utilities commissioner. Lynch was concerned about proposed regulations that would open the California gas market to LNG. Acting on her own, Lynch alerted these activists and encouraged them to be in contact and to get involved and collaborate in the upcoming California LNG proceeding. In this sense Lynch – as a "neutral expert" – played much the same role in the incipient West Coast movement as fisheries regulators did in the Gulf Coast case. Lynch's efforts brought together representatives of regional and local opposition to LNG on the West Coast, who in turn founded the umbrella organization, Ratepayers for Affordable Clean Energy (RACE) in early 2004.[20] RACE operated out of Pacific Environment and began to hire consultants and seek volunteers to develop filings for the California LNG proceedings.

At the same time, a well-known coastal advocate from Santa Barbara, Susan Jordan, was also becoming concerned about the West Coast LNG proposals from reading newspaper reports: "I became troubled . . . because I could tell [LNG] was being heavily pushed by the Bush administration. There was no needs assessment and no comprehensive planning about

[18] R. Cox, interview with second author, August 11, 2009.
[19] Baja opponents also used environmental justice frames because these projects were mainly meant to serve the California market. For more information about the Baja-based movement against LNG, see Carruthers (2007).
[20] Cox interview, August 11, 2009.

where projects should be sited . . . . It was just a free for all."[21] As a result of these concerns, Jordan purposively engaged in relational diffusion to broker connections between twenty-five different environmental organizations with whom she had previously worked. The Statewide LNG Stakeholder Working Group was forged out of these new connections. The group began meeting regularly in early 2004 with the goal of taking a "critical eye" to the push for LNG terminals in California.[22] The group lobbied Governor Schwarzenegger to conduct a LNG needs assessment and rank proposals. They also introduced analogous bills in the California State Legislature for several years starting in 2004.

At this point in the LNG permitting race on the West Coast, the Energia Costa Azul project had been approved in Mexico but faced legal challenges. In California, the Long Beach LNG Import Terminal (onshore at Long Beach) and Cabrillo Port (offshore near Oxnard and Malibu) were the two front-runners because the three other onshore proposals had been withdrawn. The Long Beach Terminal was located in the Port of Long Beach, one of the country's busiest ports. Many state officials were concerned about the potential impacts on public health and commerce that would result from placing a potentially explosive LNG facility in such a busy location where many dangerous chemicals were already transported and stored.[23] In addition, the Long Beach project became the subject of a jurisdictional dispute (along with other onshore LNG facilities around the country) between federal and state regulators. When the proposing company filed an application only with federal regulators, state regulators sued.[24] Because state officials took such an active role in the Long Beach proposal, regional LNG opponents chose to focus their efforts on Cabrillo Port.

The influx of support from regional LNG opponents to Oxnard during late 2004 transformed previously local efforts against Cabrillo Port into a

---

[21] S. Jordan, telephone interview with second author, November 19, 2007.

[22] Jordan interview, November 19, 2007.

[23] M. Schwebs, telephone interview with second author, May 18, 2006.

[24] Before the lawsuit was resolved, the U.S. Congress passed the 2005 Energy Policy Act, which granted federal regulators exclusive authority over onshore LNG applications. City of Long Beach officials and California senators (along with senators from other coastal states in the Northeast facing similar proposals) lobbied extensively to prevent the inclusion of such language in the Energy Policy Act but lost. In many ways, these efforts represent the only example in which opponents from separate regions, specifically the Northeast and California, worked together to influence federal legislation pertaining to LNG. Their lack of success could be one of the reasons anti-LNG forces never became a national movement (to be discussed in a subsequent section).

large-scale, regional campaign. Despite citywide opposition to the project from elected officials and local environmental and social justice groups (largely focused on safety concerns), Oxnard opponents had struggled to galvanize broader support for their efforts (Boudet and Ortolano 2010). Given that the key decisions regarding the proposal were to be made at the state and national levels, mobilizing broader opposition would be critical to defeating the proposal. With help from RACE, Jordan's California Coastal Protection Network, and the Sierra Club, opposition to Cabrillo Port spread to Santa Barbara and Malibu, where movie stars like Pierce Brosnan and Dick Van Dyke joined the campaign. Opponents began to plan major events, drawing more than two thousand people first to a "paddle-out" protest in Malibu and then to the Final EIS hearing in Oxnard. In addition, Jordan hired sophisticated legal expertise in the form of the Santa Barbara–based Environmental Defense Center (EDC) – a group she had previously worked with on other coastal issues – to review the technical details of the proposal. Regional opponents also expanded the framing of the opposition to Cabrillo Port from safety and terrorism to questioning the need for LNG and highlighting potential air-pollution and climate change impacts (Boudet and Ortolano 2010).

Cabrillo Port's demise began when two of three commissioners of the California State Lands Commission voted against the facility at the Final EIS hearing in April 2007. Both cited concerns about regional air pollution, the lack of a clear need for LNG, and climate change impacts as reasons behind their decisions (Cabrillo Port Final EIR Hearing 2007). This ruling served much the same certifying function in regard to the movement and issue on the West Coast as Governor Blanco's veto had in the Gulf Coast. As Cox said in our interview with him, "it became the first time that any state officials said publicly that there was no need for these facilities and discussed lifecycle greenhouse gas emissions .... [Cabrillo Port] served as a huge victory for other projects [on the West Coast]."[25] As in the Gulf Coast, in certifying the issue, the ruling set an important precedent for other public officials. Several days after the hearing, the California Coastal Commission voted against the project and in May 2007 Governor Schwarzenegger drove the final nail into the coffin when he vetoed the Cabrillo Port proposal.

By this time, the regional movement against LNG on the West Coast was well established. Subsequent offshore proposals planned in Southern California (Clearwater and OceanWay) quickly met with fierce local and

---

[25] Cox interview, August 11, 2009.

regional opposition led by many of the same groups and individuals. The speed of these responses is the best evidence that a coordinated movement was now firmly in place. According to Cox, RACE has now transferred its efforts to aid opponents in Oregon and Washington, where several communities currently face onshore LNG proposals along the Columbia River. As a result of the efforts of West Coast LNG opponents, no proposals have been approved in California and only one has been built in Baja.

Opposition to LNG on the West Coast thus provides another example of a regional movement. Opposition efforts have intensified with each succeeding proposal announced on the West Coast. For example, the number of individuals turning up at events in Oxnard grew from one hundred at a rally to prevent the sale of land for an onshore LNG facility proposed by Occidental Petroleum in May 2002 to two thousand at the Final EIS hearing on Cabrillo Port in April 2007. From the beginning, however, local mobilization against proposals on the West Coast has tended to develop quickly, usually within a month of the initial announcement. Therefore, it has been difficult for regional anti-LNG organizations to speed up this response time. In general, the process of scale shift for the West Coast movement has been more complicated and less tightly connected than efforts on the Gulf Coast, mostly because many more individuals and groups were involved and local opposition has been so strong. What is clear, however, is that brokerage was as important to the emergence of a West Coast movement as it had been on the Gulf Coast.

As in the Gulf Coast, West Coast brokers played an important role in frame bridging and coalition building. Brokers, like Lynch of the California Public Utilities Commission (CPUC) and Cox of Pacific Environment, helped opponents move beyond localized concerns about safety and land use to fashion a broader framing of LNG opposition that included concerns about need, cost, air pollution, supply chain impacts, greenhouse gas emissions, and disruption of the transition to renewable energy. This broader oppositional frame has now been deployed across multiple cases. It also has implications beyond the locally affected communities, especially considering Governor Schwarzenegger's efforts to limit greenhouse gas emissions in California. Cabrillo Port opponents, led by Jordan, began to frame the governor's support of LNG as directly contrary to his stated goal of slowing climate change through the use of renewable energy, even showing up to protest at his signing of climate change legislation in Malibu in September 2006 (Boudet and Ortolano 2010). This broader framing is also demonstrated by the 2008 Pacific Environment report "Collision Course: How Imported LNG Will

Undermine Clean Energy in California" whose cover displays a picture of an LNG tanker leveling wind turbines (Cox and Freehling 2008). Cox explained the RACE framing strategy:

As long as [lack of] need [for LNG] issue was clear, there was no need to have competition between communities facing these proposals. RACE's argument was to keep off everything LNG and that there was a lack of need. We made a convincing case, particularly with Lynch as a spokesperson and Power's analysis. The mantra for all campaigns was that we don't need it. It wasn't about one LNG versus another LNG; it was about LNG versus renewable. We framed it that way. The bigger fight was against fossil fuels, for cleaner air and against greenhouse gases.[26]

A comparison of the "top ten" reasons to oppose LNG that was offered by opponents of early and late LNG proposals shows how opposition efforts were reframed throughout the course of the debate. As shown in Table 5.2, the top ten reasons offered by opponents of the early Mare Island Energy Project includes no mentions of the larger issues of need, cost, or renewable energy and is exclusively focused on threats to Vallejo's population. Even the ninth reason about a lost opportunity for the future was not related to the transition to renewable energy but instead focused on a desire to avoid an industrial future for Vallejo. In contrast, half of the top ten reasons offered by opponents of OceanWay, which was proposed four years later near Los Angeles, are related to general environmental opposition to LNG and not localized concerns. Moreover, arguments against all LNG are even more potent in Oregon and Washington because the majority of the gas from the projects is slated for California.

Once aware of the LNG issue (generally through media reports of local opposition efforts), regional environmental groups – like Pacific Environment, Sierra Club California, EDC, California Coastal Protection Network, and Border Power Plant Working Group – were much more aggressive than their counterparts in the Gulf Coast in becoming involved in local struggles. Thus relational diffusion from local to regional chapters of existing organizations was much less important on the West Coast as compared to the Gulf Coast. For example, Cox from Pacific Environment approached almost every local community facing an LNG proposal with an offer of assistance and encouragement to join RACE. Similarly, leaders of the Border Power Plant Working Group approached local opponents in Baja, California, with an offer to provide technical information and assistance after an initial meeting with environmental leadership along the border. After the successful defeat of Cabrillo Port, the

[26] Ibid.

TABLE 5.2. *Change in Framing of LNG Opposition from Early to Late Proposal on West Coast*

Top Ten Reasons to Oppose Mare Island Energy Project	Top Ten Reasons to Oppose OceanWay Project
1. Extreme safety hazards.	1. OceanWay is economically hazardous for residents.
2. Serious health risks.	2. California simply doesn't need an LNG import terminal.
3. Damage to waterways and water quality.	3. OceanWay increases the risk of a terror attack.
4. Disruption of maritime traffic by tankers.	4. LNG is not a clean fuel.
5. Lost property values.	5. It will increase our dependency on foreign energy sources.
6. Bechtel and Shell do not make good neighbors.	6. OceanWay poses a threat to marine life.
7. Bechtel and Shell target vulnerable communities.	7. It raises risk of damage from earthquakes and other accidents.
8. Destruction of the Mare Island Reuse Plan.	8. OceanWay is not sensitive to environmental justice concerns.
9. Lost opportunity to set a new course for our future.	9. OceanWay would create a noise problem for those onshore.
10. Loss of Vallejo's only regional park (Vallejo for Community Planned Renewal 2003).	10. OceanWay undermines California's efforts to move to renewable energies (No Way On OceanWay 2008).

California Coastal Protection Network and EDC quickly transferred their efforts to advise and sometimes lead opposition efforts to subsequent proposals, particularly in Southern California, where they used the same combination of legalistic and contentious tactics that was employed to defeat Cabrillo Port.

West Coast opponents also took these regional alliances one step further than their Gulf Coast counterparts, establishing two stand-alone regional organizations – RACE and the LNG Statewide Stakeholder Working Group. These groups became sources of information and co-ordination for regional opposition activities. For example, RACE held monthly conference calls for local opposition groups up and down the West Coast and organized mobilization activities across proposals. The first such event was a Global LNG Summit in June 2004 that brought together anti-LNG activists from around the world and offered a vehicle that facilitated collaboration among West Coast activists. Many important players in the regional anti-LNG movement attended this event, including Powers, Lynch, Cox, Jordan, and leaders from local opposition groups up and down the West Coast. Groups shared experiences and tactics from their own fights. Moreover, messages against LNG more generally (e.g., LNG is not needed, LNG is not economical, LNG will deter a transition to renewable energy, and LNG creates serious climate change concerns) were reinforced. In addition, RACE used the momentum from the summit to organize an LNG Road Show in early 2005 – which brought anti-LNG messages to Baja, San Diego, Long Beach, and Oxnard – and a Clean Energy Day of Action in late 2006 – which consisted of simultaneous anti-LNG protests in Long Beach, Oxnard, North Bend, Oregon, and Longview, Washington. To this day, RACE continues the tradition of holding an annual LNG conference for West Coast opponents.

Both RACE and the LNG Statewide Stakeholder Working Group also attempted to influence legislation at the state level that related to multiple proposals. For example, RACE sued the CPUC on two different occasions concerning pro-LNG decisions.[27] Although the commission eventually decided in favor of LNG and dismissed RACE's lawsuits, RACE's effort represents a prime example of how early LNG opponents from various communities on the West Coast began to work together to influence state-level policy that was not proposal-specific. Moreover, the work leading up to the filings and lawsuits served to pull together important information about the need (or lack thereof) for LNG in California, the cost of LNG to consumers, upstream impacts from natural gas extraction, and community-level safety concerns. According to Cox, it was "a great way to build the movement and to get people [from different communities] to talk to one another and stay involved [beyond the life of single proposal]" through monthly conference calls.[28] As an example, according to Cox, Elena DuCharme [an opponent from Vallejo] stayed involved with RACE two years after the Mare Island Energy Project was withdrawn. The Statewide LNG Stakeholder Working Group also attempted to have legislation enacted throughout the course of several years at the state level that

---

[27] RACE sued in 2005 because the commission did not conduct an evidentiary hearing process prior to allowing California utilities to enter into long-term contracts for LNG (Cox 2005). RACE sued again, with support from the city of San Diego and the South Coast Air Quality Management District, because the commission voted to allow LNG suppliers to avoid treating their gas to domestic standards without conducting an environmental analysis.

[28] Cox interview, August 11, 2009.

would have required an LNG needs assessment and ranking of proposals. Although these efforts were once again initially unsuccessful, they planted the seed for Jordan's opposition to the Cabrillo Port proposal. As Jordan explained,

When I realized that no group was undertaking an environmental review of that facility, I felt like I literally had no choice ... it would be highly irresponsible to allow a facility of that magnitude to go through the [regulatory] environmental review without anybody taking a lead from the environmental side and reviewing it. That was when I made the decision to hire my attorneys – the Environmental Defense Center.[29]

Thus these efforts by the Statewide LNG Stakeholder Working Group, though unsuccessful, represent another example of how West Coast opponents attempted state-level legislation related to multiple proposals. Similar legislative efforts are now underway in Oregon.

Unlike the case of the Gulf Coast, however, regional LNG opponents on the West Coast struggled to forge certifying alliances with state and regional government officials. Until very recently, the critical votes of the two California State Lands commissioners in 2007 and Governor Schwarzenegger's veto in that same year were really the only significant exceptions to this rule. In California, many government entities, including the governor, Energy Commission, and CPUC, strongly believed (particularly after the electricity crisis during the summer of 2001) that LNG would serve as a "bridge" to a renewable energy future. Lynch, then a Public Utilities commissioner, served as one of the lone voices in California government against LNG importation, and it was her efforts to encourage public involvement in commission proceedings that originally spurred the formation of RACE. Thus we see a similar pattern to the Gulf Coast in that the initial concerns of a broker/regulator were conveyed to activists who then took up the fight. In addition, the problem of seeking out government allies was compounded by the fact that, while RACE was suing the CPUC for changing rules related to LNG importation, Long Beach opponents were allying with the commission around safety concerns related to the proposal for their port. In general, local Long Beach opponents were somewhat disconnected from the larger anti-LNG movement on the West Coast, despite the fact that they were members on paper. Bry Myown of the Long Beach Citizens for Utility Reform commented:

---

[29] Jordan interview, November 19, 2007.

I felt a little stupid during the RACE phone calls because I was bringing up issues the others didn't know anything or care about . . . . The Sierra Club would tell me to put out flyers at the local farmers market. They didn't get my community – Long Beach doesn't even have a farmers market! I also had to become an ally with Michael Peevey [President of the CPUC and a strong supporter of LNG generally] even though I didn't agree with most of his policies. I needed the resources of the Commission as a state agency to comb through the Long Beach proposal and EIR. However, that friendship put me at odds with other LNG opponents in the state who were suing them on another issue. In many ways, I felt more of a kinship with opponents facing onshore facilities in the Northeast.[30]

As proposals have moved into Oregon and Washington, opponents have developed stronger alliances with state and regional officials in these states, as demonstrated by the 2009 suit filed by Columbia Riverkeeper, Oregon, Washington, and NMFS against federal approval of the Bradwood Landing proposal. In general, the strength of the links between California anti-LNG activists and their counterparts in Oregon and Washington, coupled with the adoption in the Northwest of the same general oppositional frame that developed over time in California, attests to the ongoing existence and viability of a strong regional movement against LNG on the West Coast.

Northeast

In sharp contrast to the previous two cases, we did not find evidence of a regional anti-LNG movement in the Northeast. Instead, as we show in the following text, opposition remained overwhelmingly local and site-specific throughout the region. Nineteen LNG projects were proposed from Maine to Maryland between 2003 and 2008 (Figure 5.3).[31] The first proposals were announced in early 2003, and by mid-2004 energy companies had announced six onshore proposals in Maine, Massachusetts, and Rhode Island. Like the West Coast, every one of these proposals experienced local opposition and the main framing of the issue by local opponents against these initial onshore proposals focused on the potential safety and terrorism risks posed by the facilities, as well as local environmental and land-use concerns. Local opponents argued that LNG should be located in unpopulated areas or offshore. There is some evidence of collaboration between opposition groups facing proposals in similar locations, for example,

[30] B. Myown, interview with second author, November 4, 2008.
[31] Several LNG projects were also proposed on the Eastern Coast of Canada; however, these are not included in our analysis.

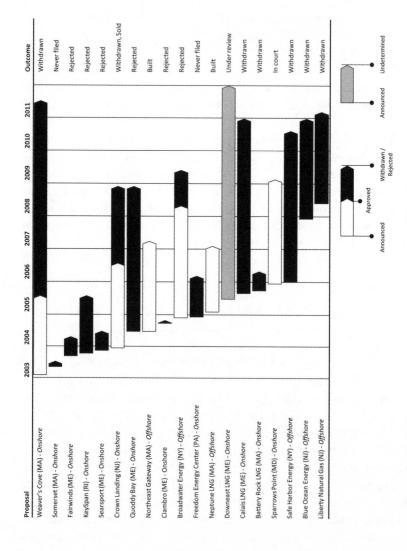

FIGURE 5.3. LNG proposals in the Northeast

opponents of Weaver's Cove and KeySpan (both would require tanker traffic through Narragansett Bay) and opponents of Quoddy Bay, Downeast LNG, and Calais LNG (all three located along Passamaquoddy Bay in Maine). However, unlike the West Coast, when energy companies began proposing offshore terminals in the Northeast during June 2004, local onshore opponents did not reframe their concerns to oppose all LNG, instead remaining firmly entrenched in localized opposition efforts and arguing in favor of the offshore option, thus preventing the development of a shared regional frame for LNG opponents in the Northeast. In addition, no regional group stepped forward to serve as a broker for anti-LNG opponents in the Northeast, as groups had done in the West Coast and Gulf Coast. Because there was no regional movement against LNG in the Northeast, we structure our narrative by location, moving from Massachusetts/Rhode Island to Maine to the Mid-Atlantic (New York, New Jersey, and Maryland).

The first onshore facility proposed in the Northeast, Weaver's Cove, was announced in February 2003 for Fall River, Massachusetts. This project experienced fierce local (and eventually state) opposition, which coalesced into the Coalition for Responsible Siting of LNG facilities in March 2004. Within a year of the Fall River proposal, LNG terminals were also announced in nearby Somerset, Massachusetts, and Providence, Rhode Island. The Somerset proposal was quickly opposed by the same groups that were against the Weaver's Cove facility and was never filed with regulators. In contrast, the KeySpan proposal, which was to be attached to an existing LNG storage facility, was initially supported by local elected officials in Providence. However, Rhode Island Attorney General Patrick Lynch and his staff, two of whom had negative past experiences with LNG, became increasingly concerned about the safety risks posed by the KeySpan facility and chose to take an active role in opposing it. The attorney general's office focused on legal means to delay and disable the KeySpan proposal. It examined federal safety regulations and gathered testimony from recognized experts such as Richard Clarke, former counterterrorism advisor to President Clinton and President Bush, and Dr. Jerry Havens, professor of chemical engineering at the University of Arkansas, whose research seeks to determine the consequences of an atmospheric release of LNG. In addition, the Conservation Law Foundation, a powerful environmental group in the Northeast, supported these activities, advocating for a regional approach to LNG and ultimately, in January 2005, coming out against the KeySpan and Weaver's Cove proposals. Although the local community response in Providence to the

KeySpan project never reached the same level of activity and organization as the response in Fall River to Weaver's Cove, the facts uncovered by the Rhode Island attorney general's office convinced many local and state elected officials to switch positions and oppose the KeySpan proposal. In June 2005, when the Weaver's Cove and KeySpan proposals came before the Federal Energy Regulatory Commission (FERC) for approval, the commissioners approved Weaver's Cove but rejected the KeySpan proposal, citing safety concerns.

With the approval of Weaver's Cove, Rhode Island opponents transferred their efforts from KeySpan to Weaver's Cove. Weaver's Cove opponents continued legislative efforts to halt the proposal. For example, U.S. Representative James McGovern (D-MA), an opponent of the proposal, inserted language into the 2005 Federal Transportation Bill preventing the use of federal dollars to demolish the Brightman Street Bridge – a bridge that Weaver's Cove LNG representatives had identified as too narrow to permit LNG tanker traffic to the facility. When the bill was approved, opponents celebrated an apparent victory, but company representatives countered by proposing the use of smaller LNG tankers that could travel under the bridge. In October 2007, the U.S. Coast Guard determined that the Taunton River (which flows into Fall River) was unsafe for LNG tanker traffic by smaller ships, and opponents celebrated another apparent victory. However, in March 2008, company representatives countered with a redesigned proposal that included an offshore berth in Mt. Hope Bay, thus avoiding tanker traffic in the Taunton River altogether. It was not until June 2011 that the company finally withdrew the proposal, citing unfavorable market conditions. As a result of the continued struggle in Fall River, LNG opponents in Providence and Fall River were unable to transfer their energy and expertise to other proposals in the region because they remain focused on defeating the Weaver's Cove facility. This undermined efforts to fashion a broader regional anti-LNG movement.

The framing by Weaver's Cove opponents has been against LNG in populated areas. Thus the announcement of an offshore LNG proposal for Gloucester, Massachusetts (Northeast Gateway, Neptune LNG), in June 2004 was actually supported by LNG opponents in Fall River. These offshore proposals were seen by many Massachusetts politicians, including Governor Mitt Romney, as a preferred option for importing LNG. The offshore proposals avoided the public-safety concerns associated with the strongly opposed onshore Weaver's Cove proposal. Thus, despite strong opposition from local fishing groups and politicians in Gloucester, mobilization against the two offshore proposals there never spread to others within the community or to environmental groups at the regional or state level. For example, despite numerous appeals from Gloucester's mayor, Fall River's mayor repeatedly voiced his support for the offshore Gloucester option as the preferred site. In addition, the Conservation Law Foundation never took on the offshore proposals directly despite requests from local Gloucester opponents to do so. Angela Sanfilippo of the Gloucester Fishermen's Wives Association lamented the fact that, after so many years spent quarreling with regional environmental groups and government regulators over fisheries protection, when those groups finally had an opportunity to work together "we lost fishing ground in the name of clean energy."[32] Once it became clear that these offshore proposals would be approved, opponents in Gloucester focused their efforts on mitigation. Coordinating an "eleventh hour" effort, they successfully secured $24 million from each proposing company, largely for local fishermen.[33] Thus, unlike on the West Coast, even local LNG opponents from the same state were unable to join forces and instead battled one another over LNG siting in Massachusetts.

In Maine, the Fairwinds LNG proposal, announced in September 2003 for Harpswell, set a precedent of local voting to determine whether a community would accept a facility. The Harpswell community voted against the Fairwinds facility in March 2004. According to the *Bangor Daily News*, "after the ugly battle in Harpswell in 2004 ... Governor Baldacci pledged that the state would not pursue siting an LNG terminal in any town that voted its opposition to such a project" (LNG Long Haul 2009). This announcement by the governor was quickly followed by votes against speculative LNG proposals in Gouldsboro and Searsport. In the communities along Passamoquoddy Bay, however, votes have been positive. For example, the Passamaquoddy Tribe voted in favor of the Quoddy Bay LNG facility at Pleasant Point in August 2004, and Robbinston residents voted in favor of the Downeast LNG project in early 2006. State and local politicians have also been supportive of these developments. Along the Passamoquoddy Bay, a coalition of residents – the Save Passamaquoddy Bay 3-Nation Alliance – has formed to oppose the facilities. In addition, these facilities have become the subject of a jurisdictional dispute between the United States and Canada. Traffic into these facilities would require tanker travel through Canadian waters, which Canadian officials, including the prime minister, oppose. As a result, all three facilities have been delayed

---

[32] A. Sanfilippo, interview with second author, July 29, 2009.
[33] J. Bell, interview with second author, July 29, 2009.

significantly in the wake of the jurisdictional controversy. As in Massachusetts, except for facilities located in similar locations, there is no evidence of coordination among local opposition groups in Maine.

Groups opposed to specific proposals in the Mid-Atlantic region show even less evidence of coordination than those in other areas of the Northeast. In the case of Crown Landing, the facility became part of a larger jurisdictional dispute between Delaware and New Jersey. BP proposed the facility in December 2003 for Logan Township, New Jersey, on the Delaware River north of Wilmington and south of Philadelphia. This location placed the pier of the facility inside Delaware's border and in direct violation of Delaware's Coastal Zone Act, which prohibits the construction of certain types of industrial facilities within the coastal zone. Delaware regulators rejected the proposal, with support from Delaware environmental groups that were longtime defenders of the Coastal Zone Act. New Jersey, with support from BP, filed suit against Delaware. The U.S. Supreme Court eventually decided in favor of Delaware, and BP withdrew the project in October 2008. Because the state of Delaware took such an active role in opposing the project to defend its jurisdiction, local mobilization was somewhat muted and legalistic in nature. For example, Delaware environment groups immediately partnered with the Environmental Law Clinic at Widener University to develop legal briefs on the issue. With a few notable exceptions, New Jersey residents and elected officials were largely supportive of the proposal because of the jobs and revenue it would bring to the area. Despite the eventual victory against Crown Landing, little of the opposition was transferred to other proposals. For example, few Delaware environmental groups were involved in the Philadelphia-led opposition to the Freedom Energy Center. This (eventually unsuccessful) proposal would have been located just up river in Philadelphia and would have required similar tanker routes along the Delaware River.

New Jersey provides a striking example of the lack of coordination among LNG opponents in the Mid-Atlantic. At the same time that the state was suing Delaware to gain jurisdictional control over the Crown Landing proposal (a move that indicated support of this proposal), New Jersey Governor Corzine expressed concern about the Safe Harbor Energy offshore proposal, 13.5 miles south of Long Beach, New York. Governor Corzine requested and was granted status as an adjacent coastal state to the Safe Harbor Energy facility by the Maritime Administration, thus providing him with veto authority. The proposing company sued over this designation but has since withdrawn the proposal.

A different offshore proposal – Broadwater Energy – has, thus far, been the most contentious proposal in the Mid-Atlantic. Located in the middle of Long Island Sound in the state waters of New York, the project spurred the formation of the Anti-Broadwater Coalition that eventually included more than fifty local civic groups and was led by the Citizens Campaign for the Environment. The coalition named the Broadwater Energy proposal "the biggest environmental threat to the [Long Island] Sound since the [Shoreham] nuclear plant" (Incantalupo 2005). In this case, New York and Connecticut politicians worked together to oppose the facility with some even kicking in funds to launch a legal battle against the proposal. According to Adrienne Esposito of the Citizens Campaign for the Environment, who led much of the opposition's effort, its main concern was the potential to set a national precedent by transferring a section of the Long Island Sound from public to private hands.[34] FERC hearings on the Draft EIS for the Broadwater Energy proposals in November 2006 drew more than two thousand attendees, and opponents gathered more than eighty thousand signatures on a petition against the proposal by December 2007. Predictably, in spite of all this opposition, the FERC approved the facility in March 2008. However, New York still had a say regarding specific permits related to the project. In April 2008, newly sworn in New York Governor David Paterson announced his opposition to the proposal, and the permits were denied. Broadwater Energy appealed the state's decisions to the U.S. Secretary of Commerce, but the state's decisions were upheld in April 2009, effectively ending the Broadwater Energy proposal.

However, despite repeated requests by anti-LNG activists opposing subsequent proposals south of Long Island and offshore New Jersey (Safe Harbor Energy, BlueOcean Energy, and Liberty Natural Gas), little of the momentum from the Anti-Broadwater campaign has been transferred to oppose other LNG proposals in the Northeast. Esposito explained her hesitancy to get involved:

We have chosen not to get involved in those [projects] because . . . Broadwater was a four year campaign for us. We poured resources into it – financial resources, staff resources – everything we had. We exhausted ourselves and almost went broke doing it. So we're not about to launch into another campaign automatically right now. We have so many other issues we're working on. And, frankly, those projects, I have been following them closely, and I don't think they're going to make it through state review either. Based on what we know now, if we need to get involved

---

[34] A. Esposito, telephone interview with second author, February 11, 2010.

for some reason, we may, but, right now, I don't see the need. They are other groups working on it and that's great.[35]

With local activists in New York and New Jersey unable to connect, it is little wonder that there is no evidence of connection between these groups and the local activists in Baltimore, Maryland, where the Sparrows Point project is currently under review. Like many other projects described in the preceding text, the Sparrows Point project is opposed locally and has become the subject of a jurisdictional dispute. In response to the proposal, in June 2006 – just seven months after the project was proposed – Baltimore County enacted zoning restrictions that would prevent an LNG facility from being located within five miles of residences. The proposing company, AES Corporation, then filed suit against the zoning law and won. Baltimore County responded by passing new legislation designed to block construction by designating the project site as an "environmentally sensitive coastline" in February 2007. AES lost initially but won on appeal in May 2008. The FERC approved the facility in January 2009, again despite obvious local opposition, and denied a rehearing request by opponents in December 2009. At the same time, the U.S. Court of Appeals upheld the Maryland Department of Environment's denial of the water-quality permit for the facility. Thus, like many other projects in the Northeast, Sparrows Point remains tied up in the state's approval process. As with virtually all of the other proposals reviewed in the preceding text, contention over the Sparrows Point project turned almost entirely on site-specific issues and localized opposition.

Anti-LNG mobilization in the Northeast thus remained fragmented and overwhelming local when compared to efforts in the West Coast and Gulf Coast. Even active opponents of specific projects would not describe LNG opposition in the Northeast as a regional movement, as demonstrated by the following statement from Esposito, who led the Anti-Broadwater Coalition:

My definition of a regional movement is when we're working together, when there is a network – even if it's a loose network – of organizations that are sharing resources, or information, or collaborating in some way. That's what I think of as a regional movement. Like, for instance, there is a regional movement in the Northeast right now to protect estuaries and we're collaborating together to increase EPA funding for estuary protection. We're working from Maine to Long Island to do that. That's a regional effort but the LNG stuff really was not that.[36]

---

[35] Esposito interview, February 11, 2010.
[36] Ibid.

Robert Godfrey of the Save Passamaquoddy Bay 3-Nation Alliance in Maine described opposition efforts in the Northeast as follows: "We're most concerned about our own individual projects, and we don't seem to connect elsewhere. I have tried to connect but there has not been a great amount of interaction, which is surprising given the fact that we're facing many of the same issues."[37]

We close by summarizing the main lines of argument that we think emerge from the Northeast narrative. Essentially, none of the key mechanisms that we saw at work in the Gulf Coast and West Coast were evident in the Northeast. Perhaps most importantly, no "neutral" expert sought to broker connections between disparate opposition groups. The few attempts at brokerage that we uncovered in our fieldwork in the region were mounted primarily by environmentalists (or other partisan actors), whose "extreme" views aroused suspicion among potential coalition partners. In all of these cases, then, the attempt to forge connections and put the movement on a broader regional footing came to naught.

Owing perhaps in part to the absence of the kind of "neutral" brokers we saw in the other two regions – figures who were able not only to link previously disconnected groups but also to fashion shared frames in order to bridge their differences – no consistent framing of the issue emerged across multiple projects in the Northeast. For example, framing of opposition to the Massachusetts and Rhode Island onshore proposals largely focused on the potential safety impacts of tanker traffic and thus called for "responsible siting" of these facilities away from populated areas. However, the main framing of opposition to the offshore Massachusetts projects centered on fisheries impacts. In Maine, opponents voiced safety and environmental concerns, while in the Mid-Atlantic, much of the opposition focused on safety and jurisdictional issues. The lack of a regional broker has clearly contributed to the absence of an overarching opposition frame across the Northeast. As we have seen, no regulatory officials or moderate environmental groups have seemed willing to fill this void, as they did in the West Coast and Gulf Coast. The important implication here is that brokerage cannot be accomplished by any entity but requires the right individual or group to be successful, probably an actor less vested in opposition to a specific project and with a geographically broader, and seemingly less partisan, view of the issue at stake.

---

[37] R. Godfrey, telephone interview with second author, January 6, 2010.

Finally, no influential public official stepped up and certified the issue in a decisive and regionally salient way.[38] To be clear, there were cases in which prominent officials came out in opposition to a particular LNG project, but in all of these instances, the opposition was seen as peculiar to the case in question. So the governor of Delaware opposed the Crown Landing project slated for the opposite shore of the Delaware River in neighboring New Jersey. But her opposition was less a matter of principled opposition to LNG than a function of an arcane jurisdictional dispute involving state boundaries set by an obscure nineteenth-century legal ruling. As a result her opposition was seen as having little relevance for other cases within the region.

Beyond the absence of these mechanisms, however, it is important to acknowledge two important contextual factors that probably also impeded the development of a coordinated movement within the region. The first is simply the fact that the projects proposed for the region were located in so many different states, each characterized by a different regulatory regime. These varying regimes and partisan political realities made the lessons learned and tactics employed by opponents in one state much less relevant to potential activists in others. For example, affected communities in Maine were allowed to vote on whether to accept a facility. Influencing a community vote would suggest a very different oppositional strategy than, for example, those communities along the Delaware-New Jersey border that were faced with the Crown Landing proposal. In this case, local votes would not matter, and instead opposition required significant legal knowledge because the project became the subject of a jurisdictional dispute that eventually reached all the way to the Supreme Court. In contrast, on the West Coast and the Gulf Coast, where multiple projects were proposed for most states, lessons learned and the momentum from one project could easily be transferred to another because decision-making structures and thus oppositional strategies remained more constant and relevant across multiple sites.

The second contextual factor is the Northeast's dependence on coal for electricity generation and oil for residential heating (especially when compared with the West Coast's energy-supply mix). We will have more to say about this later in this chapter, but the prospect of a cleaner-burning supply of natural gas from LNG meant regional regulators and

---

[38] This finding closely parallels results reported in Sherman's (2011a) comparative study of variable opposition to low-level radioactive waste disposal in the United States between 1979 and 1999.

environmental groups, like the Conservation Law Foundation, were loathe to disparage all LNG proposals. Instead, these groups chose to evaluate each proposal on an individual basis, thus preventing them from serving the all-important brokerage or certification role for the development of regional anti-LNG movement as in the West Coast and Gulf Coast.

WHY NO NATIONAL ANTI-LNG MOVEMENT?

Figure 5.4 shows the regional nature of the anti-LNG movement. It provides a breakdown of newspaper articles about LNG by region. As LNG proposals spread from the Gulf Coast to the Northeast to the West Coast (first in California, with a peak in coverage in 2005, and then in Oregon, with a peak in coverage in 2007–8), so did coverage and opposition. With each successive region, media coverage became more extensive. However, coverage in national newspapers peaked in 2005 and has since diminished. This decrease in national coverage occurred despite regional opposition to the four LNG proposals then being considered in Oregon – opposition that forced both Democratic presidential candidates to take positions on LNG siting during the party's primary in 2008. Similarly, e-mails by members of a national Listserv on LNG safety also peaked in 2005 and have since decreased, as shown in Figure 5.5.

These two figures suggest that there was a moment in time, especially during the summer of 2005, when the anti-LNG movement looked like it might go national in scope. Yet, according to movement activists, who have a vested interest in exaggerating the size and strength of any such movement, anti-LNG activity remained exclusively regional. Specifically, within each region, there were organizational forces at work to oppose LNG projects. However, this opposition never crystallized into a national struggle against LNG in the way that, say, the antinuclear movement did during the 1970s.

The question is, why not? The answer should be clear, we think, from our analysis of the way LNG proposals were received in the three regions that we have examined. Quite simply, the lack of coordinated opposition within the Northeast made the question of a national anti-LNG movement moot. You can hardly have a national movement with such a large and populace part of the country missing in action. More generally, the disparate reactions to LNG within each region – for example, narrow opposition to open-loop systems in the Gulf Coast and generalized opposition to LNG in the West Coast – virtually ruled out the possibility of a true national movement developing around the issue. Although there were some attempts at national-level action

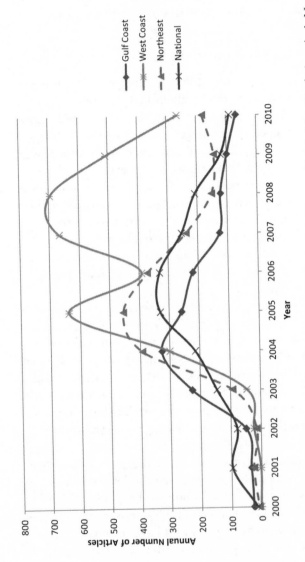

FIGURE 5.4. Newspaper coverage of LNG by region. This figure was created by searching in NewsBank-America's Newspapers for articles containing the phrase "liquefied natural gas." For each region, the top four newspapers in terms of coverage were included in the total. For the Gulf Coast, the newspapers included were the *Houston Chronicle* (Texas), *Mobile Press-Register* (Alabama), *Beaumont Enterprise* (Texas), and *Midland Reporter-Telegram* (Texas). For the West Coast, the newspapers included were the *Daily Astorian* (Oregon), *Ventura County Star* (California), *Coos Bay World* (Oregon), and *Long Beach Press-Telegram* (California). For the Northeast, the newspapers included were the *Fall River Herald News* (Massachusetts), *Bangor Daily News* (Maine), *Boston Globe* (Massachusetts), and *Portland Press Herald* (Maine). For national coverage, the newspapers included were the *Washington Post*, *New York Times*, *Congressional Quarterly Transcriptions*, and *Associated Press Archive*. Data is as of May 8, 2011.

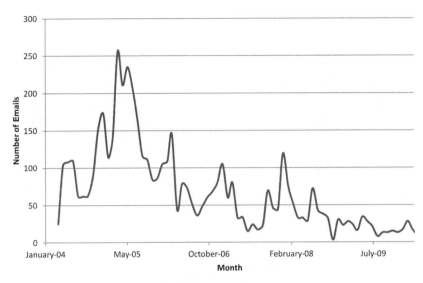

FIGURE 5.5. E-mails sent on national LNG safety Listserv over time. This chart was created using data available through the LNG safety Listserv on Yahoo! Groups.

(through the *CPUC vs. FERC* lawsuit, which was supported by U.S. congressional representatives from several impacted states and by amendments offered to the 2005 Energy Policy Act in an effort to limit FERC jurisdiction over siting issues) and global-level action (through the Global LNG Summit in June 2004 in San Diego), nothing resembling a national movement against LNG has, or is likely, to develop for the foreseeable future.

There is one other obvious constraint on the development of a true national anti-LNG movement within the United States. In contrast to other similar issues (e.g., nuclear power plants, hazardous waste disposal sites, and incinerators), the requirement that LNG terminals be located along the coast renders the issue irrelevant for the territorial majority of the country. Setting aside this obvious constraint, we argue that there are four reasons why such a national movement did not materialize: lack of a cross-cutting frame, lack of a national broker, an inhospitable national political environment, and industry learning.

Incompatible Collective Action Frames

Neither of the frames selected by the two successful regional anti-LNG movements in the Gulf Coast and West Coast could have served as the

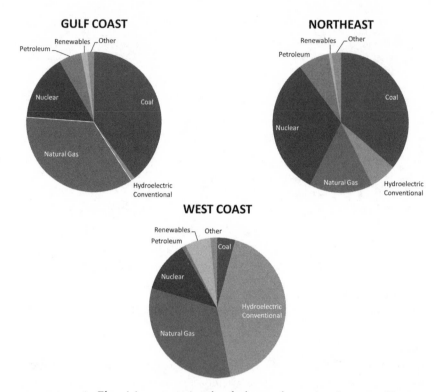

FIGURE 5.6. Electricity generation by fuel type by region in 2000 (Energy Information Administration)

substantive basis for a national anti-LNG movement. The Gulf Coast's stance against offshore, open-loop LNG was not relevant in other parts of the country because no facilities of this type were proposed elsewhere. The West Coast's stance against all LNG was not suitable to the other two regions, where at least some state officials and even some environmentalists were anxious to replace coal-fired electricity generation with cleaner-burning natural gas.[39] Figure 5.6, which graphs electricity generation by fuel type in each region during 2000 (when LNG proposals were beginning to be announced), shows how much more dependent on coal the Gulf Coast and Northeast regions were when compared to the West Coast. During interviews in the Northeast and Gulf Coast,

---

[39] In addition, interviews of Gulf Coast opponents made it clear that opposition to all LNG development would have been culturally unacceptable because many onshore facilities were supported by the locally affected communities.

opponents were clear that their opposition was not against all LNG but instead against specific types of projects and specific locations. For example, opponents in Mobile, Alabama, recently endorsed a closed-loop version of the Bienville Port proposal, which would be located south of Compass Port. Casi Callaway of Mobile Baykeeper explains, "The reason we would not fight all LNG is because we're not California and we burn [a great deal of] coal. Coal is awful and it's the most polluting thing in the whole wide world. We even have some gas turbines at some of our power plants that don't get used because gas is too expensive .... We wanted additional sources of natural gas."[40] The Conservation Law Foundation, a major environmental group in the Northeast, maintains the following position on LNG terminal siting:

Natural gas has important environmental benefits. Burning natural gas, instead of oil and coal, to create electricity results in less harmful air pollution and fewer emissions of greenhouse gases that cause global warming. While the region should take advantage of every opportunity to reduce overall energy demand through increased efficiency and reduce demand for fossil fuels through increased use of renewable energy, the Conservation Law Foundation considers LNG to be an important transitional fuel until the region moves to a comprehensive renewable energy base .... The Conservation Law Foundation has called on the FERC, New England's coastal governors and other key policymakers to develop a regional strategy for siting LNG terminals in New England. (Conservation Law Foundation 2010)

Esposito, of the Anti-Broadwater Coalition, also comments on the difference in framing among anti-LNG groups,

All through this [experience of opposing Broadwater], we were contacted by other groups. We were contacted by the groups in the Northeast; we were contacted by groups in California; we were contacted by groups in Louisiana; we were contacted by groups in Rhode Island and some as far away as Canada. But, you know, some of those groups were just anti-LNG, but that was not our position. Our position was estuary protection. So some of the groups were not overly thrilled about that, but that's our position.[41]

Although this stance against specific types and locations of LNG terminals and for a regional siting strategy was more prevalent for early West Coast projects, like the Mare Island Energy Project, opponents of subsequent West Coast projects, like Cabrillo Port and Long Beach, were much more sweeping in their critique of LNG.

[40] C. Callaway, interview with second author, April 9, 2009.
[41] Esposito interview, February 11, 2010.

Absence of a National Broker

This lack of a consistent frame across regions meant that brokerage was difficult, or conversely, the lack of a national broker(s) made consistent framing difficult. We are aware of separate attempts by two environmental organizations to take a national position against LNG – the Sierra Club and Waterkeeper Alliance. Both were unsuccessful largely because representatives from the Northeast and Gulf Coast argued against a blanket rejection of these types of facilities. For this reason, Sierra Club chapters in California and Oregon have listed stopping LNG terminal siting as one of their priorities (Sierra Club California 2005; Sierra Club Oregon 2010), while Sierra Club National maintains the following policy:

The role of natural gas in the transition to a clean energy future varies by region. The Sierra Club supports regional planning to develop future scenarios for natural gas use and assess all available supply and efficiency resources using full-cost accounting, including environmental costs and benefits. Such planning on a regional basis is needed to assess the size and timing of necessary new gas resources, if any, whether from domestic or LNG supply, as well as investment in cost-effective efficiency resources. Broad positions by Sierra Club entities, including chapters, concerning LNG must be based on in-depth regional analysis and include the participation of all affected chapters. (Sierra Club 2009)

Likewise, Callaway of Mobile Baykeeper vividly recalls arguing side by side with Waterkeeper Alliance members from Connecticut against California representatives in an effort to avoid a blanket anti-LNG stance at the Annual Waterkeeper Alliance meeting:

I've had some conversations with Californians, because we're part of the Waterkeeper family, and I had to make sure that our national policy was not absolutely no – so that decisions could be made on a local basis. The Connecticut and other Northeast folks are also not opposed to LNG, just very opposed to the wrong siting. We're trying to be balanced.[42]

Despite the experience of Callaway at the Waterkeeper Alliance meeting, at least among some West Coast opponents, there has been an unwillingness to foist anti-LNG sentiment upon other more coal-dependent regions. Cox of Pacific Environment and RACE explains,

We limited this [anti-LNG] campaign to the West Coast, and, unless specifically asked, we didn't do much on the Gulf or East Coasts, mostly because they are so coal dependent. California and the West are not coal dependent. I don't know what

---

[42] Callaway interview, April 9, 2009.

it's like to have a coal-fired power plant in my neighborhood and I don't know how I would feel about liquefied natural gas if I did. Also, in California, we have these big sweeping, ambitious … renewable laws. If we have plans for a renewable future, why are we talking about LNG?[43]

Thus, because of the varying stances among regional anti-LNG activists, brokering an alliance among them – even within national environmental organizations, perhaps the most likely ally – proved impossible. Opposition to domestic shale gas development seems likely to follow a similar path if news reports are any indication. Despite fierce opposition to this type of development from the New York chapter of the Sierra Club, the position of the National Sierra Club, Environmental Defense Fund, and Natural Resources Defense Council remains favorable, given appropriate regulation, because of natural gas's ability to replace coal as a cleaner-burning fuel for electricity generation (Casselman 2009). In many ways, domestic shale gas is preferable to LNG because it does not require foreign supplies or long-distance tanker travel, making blanket opposition even more problematic.

Inhospitable National Political Environment

Thirty years of social movement scholarship has affirmed the important role that favorable "political opportunities" often play in shaping the emergence and development of contentious politics (for a summary of work in this tradition, see Kriesi 2004). A particular dimension of political opportunity figured prominently in our accounts of the emerging regional movements in the Gulf Coast and West Coast. In both regions, opposition forces benefited enormously from the proactive efforts of elite allies. These allies took two forms. First "neutral experts" in state agencies in both regions not only helped to broker connections between disparate opposition groups but also to fashion the broader collective action frames that harmonized their views on the issue. Second, state officials – led by Louisiana Governor Blanco in the Gulf Coast and two state land commissioners and later Governor Schwarzenegger in California – certified the movement at a critical juncture in its development, energizing opponents and legitimating the movement in the eyes of other politicians in the two regions.

At the national level, however, the situation was very different and much less hospitable to movement forces. The Bush administration was

---

[43] Cox interview, August 11, 2009.

highly supportive of LNG development. It was Bush's election in 2000 that triggered the rash of LNG proposals in the first place! (This encourages the following theoretical aside: the notion of a political opportunity structure does not apply to movement groups alone. All strategic political actors – including, in this case, big oil and gas – can be expected to respond to perceived shifts in political opportunities. In this instance, Bush's election was seen as a boon by the gas and oil industry, triggering a great deal of strategic initiatives, including the LNG proposals that flooded the country beginning in the year – 2001 – that Bush took office.) Needless to say, the Bush administration did not disappoint. Fulfilling a campaign promise, less than four months after taking office, Bush issued Executive Order 13212, "Actions To [*sic*] Expedite Energy-Related Projects." The intent of the order was clear from its title; Bush aimed to "expedite" energy development and streamline the review of energy projects. The message to federal agencies charged with reviewing energy proposals could not have been clearer. Quoting the order directly, "For all energy-related projects, agencies shall expedite their review of permits or take other actions as necessary to accelerate the completion of such projects." There was a new marshal in town, and he was gunning for those who would dare to obstruct the development of domestic gas and oil.

This would remain a consistent policy emphasis throughout Bush's years in office. In 2002, the Maritime Transportation Security Act dramatically expanded the basis for offshore oil projects. In the Energy Policy Act of 2005, Bush and his congressional allies sought to limit local and state jurisdiction over onshore facilities. Repeated attempts by legislators in affected coastal states to strike LNG-related provisions in this legislation were defeated in Congress. Thus, there was little, if any, hope for national-level progress on the LNG issue. One could argue that Bush's election and subsequent executive order was the catalytic political opportunity that triggered an elite movement by oil and gas companies to propose and push for LNG facilities in the first place (Whitmore, Baxter, and Laska 2008).

Industry Learning

A final component that limited the possibility of a national anti-LNG movement was that companies proposing these facilities modified their proposals based on community response to previous proposals – that is, the industry learned from experience. For example, when onshore proposals were quickly opposed in California and the Northeast on the basis of local safety and environmental concerns, proponents went offshore.

When open-loop facilities became untenable in the Gulf Coast, companies began proposing closed-loop facilities. This proponent perspective is often ignored or overly simplified in the literature on antitechnology movements and is similar to the state perspective in rights-based movements, which is also often ignored or overly simplified by social movement scholars. However, as we note here, this proponent perspective and response remains an important consideration because of the fundamentally two-sided nature of siting debates.

CONCLUSION

We began this chapter by reviewing the few fragmented literatures that have had anything to say about the neglected topic of movement expansion (and contraction). From that exercise in conceptual parsing, we extracted three mechanisms thought to shape the process of upward "scale shift." These were frame bridging, from the NIMBY literature, and brokerage and relational diffusion from the theoretical work on scale shift (Tarrow and McAdam 2005). Based on our reading of events on the West Coast and especially the Gulf Coast, we added one additional mechanism – certification – to our list. Drawing on extensive fieldwork, we then constructed detailed analytic narratives describing the varied responses to LNG in our three regions. In bringing the chapter to a close, we very briefly revisit the role that these mechanisms played in the emergence of regional movements in the Gulf Coast and West Coast and close with some final thoughts on certification, the one variable whose link to scale shift emerged out of this research.

Based on what we observed in the Gulf Coast and West Coast, it is virtually impossible to think of frame bridging and brokerage as separate mechanisms. It might be more accurate to think of the two as constituting a linked process of coalition building, rather than as two separate mechanisms. Whether the two always, or almost always, go together is impossible to say, but it is clear that they did in the two cases on offer here. The specific dynamics involved in the deployment of these linked mechanisms were eerily similar in both regions. In both cases, state regulators brokered connections between disparate opposition groups while working to develop broader oppositional frames to cement the embryonic coalition. This was especially consequential in the Gulf Coast, where the ideological and cultural differences between environmentalists and fishermen made for strange bedfellows indeed. Were we to assign paramount importance to any of our mechanisms in regard to these two cases, the linked efforts at brokerage and frame bridging would have to get the nod.

In contrast, relational diffusion figured much less prominently in our narratives than the other three mechanisms. It was, however, important, though it was much more so in the Gulf Coast than the West Coast. There we found clear evidence of escalating mobilization and the spread of opposition within the established networks that defined the three communities – environmentalists, sport fishermen, and commercial fishermen – that came to comprise the broader movement. The sheer size, geographic breadth, and fragmented character of the movement on the West Coast made it more difficult to see the process of relational diffusion clearly. We suspect that more of this went on then we were able to detect, but we are also prepared to believe that this mechanism was considerably less important on the West Coast than in the Gulf Coast.

Certification is the last of our mechanisms. To be honest, it was not on our radar at the beginning of the study. But through our research we have come to see it as very important to scale shift, especially in the case of the kind of technical land-use movements that have occupied our attention here. Perhaps because of its origins during the 1960s and 1970s, social movement theory has been shaped primarily by a close interrogation of such "rights" movements as African American civil rights, feminism, prolife/prochoice, and so forth. In most such cases, the issue at stake already enjoys widespread legitimacy within the aggrieved population. As a result, movement expansion does not depend on certification by state actors or other influential figures. In the case of emergent, often highly technical, land-use issues, however, we are talking about a very different matter. Consider the case of the Gulf Coast. Given the regions long and generally cozy relationship with the oil and gas industry (at least until the Gulf Coast oil spill!), it is hardly surprising that the initial response to LNG – even open loop – was positive. In short, at the outset, the issue had virtually no standing with either the general public or policy makers in the region. In such cases, which we take to be broadly modal for emerging land-use issues, certification by "neutral experts" or influential public figures becomes key to the emergence and spread of an opposition movement. We amassed a great deal of evidence in support of this conclusion through our fieldwork in all three regions. Movement scholars will thus want to add certification to the list of mechanisms thought to shape the prospects for scale shift. More generally, we hope movement analysts will devote more systematic attention to the issue of movement expansion (and contraction). Analyzing the emergence of localized contention is one thing; understanding how and why that localized contention does or does not spread is quite another.

6

Back to the Future

Returning to a Copernican Approach to the Study of Contention

Having worked our way through the main empirical findings from the research, we want to use this final chapter to do three things. We begin by revisiting the four key research questions with which we opened the book, answering each in turn based on what we have learned from our investigation. But the book was never conceived as simply a conventional research monograph. Instead, the study was motivated by a broader theoretical agenda that was reflected, as well, in our research design. Concerned that the characteristic narrowness and movement-centric focus of most scholarship in the field has seriously distorted our understanding of contentious politics, we undertook a research project that we hoped would put movements "in their place." That is, instead of selecting successful movements for study, we chose to research communities – communities "at risk" for mobilization by virtue of their shared exposure to the "threat" of environmentally risky energy projects. Our goal was to downplay the emphasis on movements and focus on the ways that local context shapes the prospects for emergent contention, whatever form that takes. We also wanted to focus less on opposition groups and more on the broader array of actors who shape episodes of contention and stability and change in public policy. Consistent with this aim, our second goal for the chapter is to discuss our results in light of the substantial body of work on the policy process. This is all by way of urging movement scholars to take account of other literatures that clearly bear on their subject matter. Because we were simultaneously studying the dynamics of local conflict and, in the case of LNG, what we see as the emergence and consolidation of a new policy subsystem in the area of energy, it is important for us to tease out the implications of our work for this important literature. All of this is in the service of our general aim of contributing to a more

Copernican understanding of contention at the expense of the Ptolemaic perspective that has come, in our view, to characterize the field. Our final goal for the chapter, then, is to use what we see as the important implications of our research to call for a significant reorientation and broadening of the field of social movement studies.

REVISITING OUR QUESTIONS

At the close of Chapter 1, we laid out our empirical agenda for the book. It took the form of the following four research questions:

1. How much oppositional mobilization do we see across our twenty communities?
2. What "causal conditions" explain variation in the level of mobilization in these communities?
3. Net of other factors, what influence, if any, does the level of mobilized opposition have on the outcome of the proposed projects?
4. Why did opposition to one kind of energy project – liquefied natural gas (LNG) terminals – grow into broader regional movements in some parts of the country but not others? And what mechanisms appear to shape this upward "scale shift?"

We take these questions in turn, briefly summarizing the main empirical "punch lines" from the research.

So How Much Oppositional Activity Did We Actually Find?

As reported in Chapter 3, not very much. The summary statistics tell the story:

- By our operational criteria, only ten of our twenty communities "mobilized" in opposition to the proposed projects.
- Only half of our communities experienced at least one – and often only one – protest event.
- The mean number of protests across our twenty communities was a paltry 1.4 per locale.
- Not a single instance of disruptive protest – that is one featuring arrest, injury, or property damage – was recorded for any of our communities.
- Nor was it a case of citizens eschewing protest for more conventional, institutionalized forms of claims making. The median number

of letters to the editor regarding the proposed projects was just 5.5 across all cases.
- Only one true social movement emerged in response to the proposed projects.

What makes these numbers all the more remarkable is the fact that they were gathered, not in a random sample of all communities, but rather in locales subject to the objective risks associated with the proposed projects; that is to say in communities where, based on the received wisdom of thirty years of social movement scholarship, we might have expected to see considerable contention. It is hard, to say the least, to square these numbers with the idea that we are now living in a "movement society" (Meyer and Tarrow 1998). Our figures jive well with the very low number of protests reported by Verba, Schlotzman, and Brady in their 1995 survey of "civic voluntarism in American politics." In general, we contend that the substantive thrust of the literature – reinforcing a 1960s image of movements – combined with the methodological convention of studying movements, rather than mobilization attempts or populations at risk for contention, has dramatically exaggerated the frequency and causal significance of true social movements. Although not quite ready to say that our results should be read as a kind of "baseline" of routine contention, we would certainly argue that it is closer to that imagined baseline than the wildly nonrepresentative studies of various movements that comprise much of the empirical work conducted to date.

What "Causal Conditions" Explain Variation in Opposition across Our Cases?

In Chapter 3 we report a number of causal "recipes" that predict variation in mobilized opposition across our cases. We will not attempt to revisit those recipes here. Instead, we offer a more holistic summary of those results. As we note in Chapter 3, we see our results as broadly consistent with the logic of the political process model as originally deployed by the first author in his 1982 study of the civil rights movement. Unlike the stylized caricature of political process that is now widely accepted in the field, the original formulation of the theory did not privilege "political opportunities" as the key to movement emergence. In point of fact, there was no "first among equals" among the three factors stressed in the model. The causal logic underlying the perspective was conjunctural; movements tended to develop when all three factors – expanding political opportunities, grassroots mobilizing structures, and cognitive liberation – came together. Moreover, the causal force of the two structural components of the model – political opportunities and mobilizing structure – was held to work *through* the key subjective/ cultural process of cognitive liberation. Improving political conditions were thought to increase the likelihood of movement activity by increasing the sense of efficacy among the aggrieved. Similarly, mobilizing structures aided movement emergence in part by affording would-be insurgents the "free spaces" within which the critical processes of oppositional framing – for example, social construction and collective attribution – could occur.

We have gone to the trouble to clarify the original perspective, not because we care to revisit the tiresome polemical debate that has come to be associated with the theory, but simply to frame the results reported in Chapter 3. Bottom line: we see those results as entirely consistent with (a) the conjunctural logic of the original theory and (b) its central emphasis on collective meaning making as the ultimate catalyst of emergent contention. The recipes reported in Chapter 3 speak to both of these points. To the first of these points, it should be enough simply to note that recipes, by design, reflect Ragin's conjunctural approach to causation. Instead of discrete variables, making their individual contributions to the overall explanatory power of a model, Ragin sees causality as inextricably bound up with particular combinations of causal conditions. But what of the specific combinations reported in Chapter 3 as accounting for variation in mobilization across our cases? We reproduce those recipes in Figure 6.1.

We see these combinations as very much in accord with the initial formulation of the political process model. How so? For starters, the causal conditions "political opportunity" and "civic capacity" figure prominently in several of the recipes shown in the figure. But they tend to work in combination with – and probably through – what we have termed our "community context" variables. These include such causal conditions as "similar industry," "economic hardship," and "experience with a past proposal." Although we interpret political opportunity and civic capacity as crudely measuring the objective structural potential for oppositional mobilization, we see the community context variables as powerfully shaping the subjective interpretations of the projects by local residents. So, for example, in locales facing severe economic hardship, we expect that even risky projects are likely to be perceived as a boon to the community. If a community has a long history with a given industry ("similar industry") they are likely to be inured to the risks of a new project associated with that industry. By contrast, communities that have mobilized against similar past projects are almost certain to regard new proposals as equally

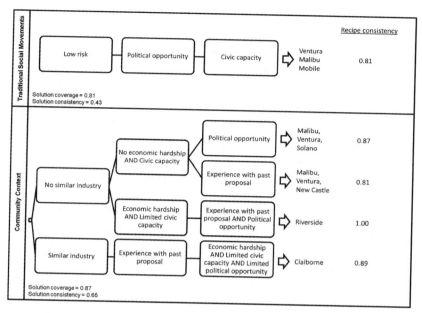

FIGURE 6.1. Mobilization recipes

objectionable. In short, what we see reflected in these recipes is very much the same blend of opportunity, organizational capacity, and locally constructed perceptions (in this case of risk) featured in the political process model. Moreover, the dynamic blending of these factors also comes through clearly, we think, in the narratives with which we close Chapter 3.

What Impact Does Mobilization – or Lack Thereof – Have on Project Outcome?

Our research yielded two significant empirical surprises. The first was the exceedingly modest levels of mobilized opposition touched on in the previous section. Based on just how little was going on in most of our communities, we quite reasonably assumed that opposition activity would have little or no impact on the fate of the projects. That led to surprise number two: variation in community mobilization appears to clearly impact the outcome of our projects. This conclusion rests on fuzzy set/ Qualitative Comparative Analysis (fs/QCA) analyses of three specific outcome conditions – approval, rejection, and built. The results of these analyses can be summarized succinctly:

- The absence of mobilized opposition is more or less sufficient to explain project approval.
- A lack of mobilization also comes close to being a necessary condition for built, with a consistency of 0.75.
- Although neither necessary nor sufficient to explain rejection, the presence of a mobilized opposition figures prominently in the recipe that does. We will have more to say on this score in the following text.

Bottom line: the ultimate fate of our projects clearly bears the imprint of mobilization or its absence.

How should we interpret these results? It is tempting to regard the findings as just that much more evidence attesting to the social change impact of social movements, to add our study to the growing list of published works that appear to document the role of movements in shaping a broad range of social, political, and cultural outcomes. In the end, we are prepared to accept this interpretation, but only after first voicing an important caution. The caution has to do, not with those who mobilized, but with those against whom they mobilized. To properly assign responsibility for the outcome of the projects, it is important to understand how the sponsoring companies regarded these initiatives. This is most clearly seen in the case of the LNG proposals that make up the great majority of our sample. It is not much of an exaggeration to say that George W. Bush's election in 2000 triggered a feeding frenzy among gas and oil companies. Determined to take advantage of the administration's favorable energy policies, these companies competed to propose LNG facilities for a majority all of the feasible sites on all three coasts, Atlantic, Pacific, and Gulf. In all, nearly fifty LNG projects were proposed between 2001 and 2005. Industry experts, however, estimated that the demand for LNG would support at most ten terminals. Thus the companies came to regard these proposals not as individually critical, but as so many irons in the fire. The strategic imperative became "tending" each "iron" as long as it made economic sense to do so. If any given iron came to be seen as too expensive to tend, the companies were perfectly willing to table, withdraw, or generally "divest" in that particular proposal. Among the developments sure to raise the specter of delays and cost overruns was any hint of opposition to a given project. Put simply, the issue of "negative project outcomes" is a tricky one in this case. The nonzero sum, irons-in-the-fire logic of the multiple proposals they were tending encouraged companies to accept negative outcomes if the economics of the situation looked problematic.

The fs/QCA analysis we conducted to explain built yielded results wholly consistent with this interpretation. Hoping to capture the impact of the kind of economic calculus described in the preceding text, we incorporated two additional causal conditions into a three-variable model that included mobilization. We called these two economic measures *regional saturation* and, appropriately enough, *irons in the fire*. The first was designed to capture the number of terminals – operated by rival companies – that were already on line or under construction in the same region as the proposal in question. A higher score on the measure meant more terminals in the region and therefore more challenging market conditions for the project under consideration. By contrast, irons in the fire was based on the number of the companies own approved projects at the time the proposal in question was approved. Again, we assumed that a higher score on this measure would reflect a much weaker motivation on the part of the company to carry the project forward beyond approval. Having already secured approval to build at least one other terminal in the same region, we reasoned that the company's enthusiasm for constructing a second facility would have declined significantly, especially if the region was also "saturated" with terminals built by industry rivals. Happily, our analysis yielded results consistent with these expectations. The recipe of low mobilization, little regional saturation, and few or no irons in the fire does a good job of explaining the outcome built.

This allows us to return to an important general point that we have stressed throughout the book. When scholars focus only on movement actors, they are bound – in Ptolemaic fashion – to overstate the significance of their actions. Our project outcome results afford an interesting case in point. Our results clearly suggest that even low levels of mobilized opposition contribute to what, from the companies' point of view, would be seen as negative project outcomes. But to really understand these results and properly apportion "credit" for the negative outcomes, it is important to understand the economic logic that the oil and gas companies brought to the LNG proposal process. Had each proposal been viewed as critically important in its own right, we are quite certain that the companies would have resisted divestment and fought that much harder to get them built. Even acknowledging the important role that the companies' assessment of prohibitive costs had on the outcome, however, we find it very interesting that even low levels of opposition activity appear to have been enough to encourage this conclusion. If private firms and perhaps government actors are this risk averse and sensitive to popular opposition, maybe we are living in a movement society after all, or more accurately, a "contentious society," because one of our central findings is that movements remain rare events, while contention – of varying form and levels – is fairly common in society.

Why Do Some Movements Scale Up While Others Do Not?

As we noted at the start of Chapter 5, the topics addressed in Chapters 3 and 4 – the emergence and impact of collective action, respectively – have been central to social movement scholarship for a long time. Scale shift – the geographic expansion (or contraction) of a movement – has been given short shrift by movement analysts. Although we certainly couldn't have known it at the outset of the study, our research afforded us the perfect opportunity to study the dynamic mechanisms that facilitate the spread of contention. As it turned out, the time period of our study coincided with the flood of LNG proposals following Bush's election in 2000 as well as the gradual consolidation of regional movements against LNG on the Gulf Coast and West Coast. Curiously, however, in spite of a similar spate of proposals targeting the Northeast, opposition to LNG remained over-whelmingly local within the region. Finally, adding yet another interesting wrinkle to the story, even as opposition to LNG spread along the Gulf Coast and out west, these two regional movements failed to become one. These developments thus granted us a unique opportunity to study scale shift in comparative perspective.

Given the paucity of systematic work on the topic, we would caution against regarding our findings as being in any way definitive. Much more work needs to be done in this area – and with a broader array of methods – to see if the mechanisms that appear to have shaped our cases recur in others. Notwithstanding this caution, we are quite confident in our narra-tive analysis of events in these three regions and, in particular, in assigning a significant causal role to the following four mechanisms: frame bridging, brokerage, relational diffusion, and certification. We see a very similar story unfolding in the Gulf Coast and West Coast. In both cases, the story begins with "neutral" state officials catalyzing and consolidating opposi-tion in their region by combining the critical roles of broker and frame bridger. In the Gulf Coast, this role was played by state and federal fisheries regulators. Not only did these regulators aggressively broker connections between fishermen and environmentalists, but they also framed opposition to "open-loop" LNG in such a way as to integrate the concerns of both groups, groups whose natural antipathy would have ordinarily made the idea of coalition unthinkable.

On the West Coast a commissioner with the California Public Utilities Commission (CPUC), Loretta Lynch, played a virtually identical role in the incipient movement there. Concerned about proposed regulations that would open up the California market to LNG, Lynch contacted local but previously disconnected environmental activists in Northern and Southern California and urged them to combine forces and begin working together on the issue. In brokering these connections, Lynch also framed the issue in the broadest possible way in order to accommodate the specific concerns motivating opponents in various locales. As we said at the close of Chapter 5, so close was the connection between brokerage and frame bridging in these two cases that we are left to wonder whether we are really talking about a linked process of coalition formation, rather than the two discrete mechanisms of brokerage and frame bridging.

The early efforts of these "neutral" officials also served a critically important "certifying" function. We speculate that certification may be more important to scale shift in the case of emergent mobilization in relation to local land-use issues than it has been in more traditional rights-based movements. The point is, in the latter, a natural constituency for the movement typically exists. The majority of the beneficiary constituents do not have to be convinced of the legitimacy, or rightness, of their cause. Ordinarily, however, this situation does not apply in the case of emerging land-use issues, especially those of a highly technical nature. Citizens are typically unaware and confused when they first confront such issues. In such situations, certification by a neutral expert or influential politician may be necessary for opposition to spread beyond isolated local pockets. The movements in the Gulf Coast and West Coast benefited from early, well-publicized instances of certification.

Finally, relational diffusion, the spread of contention along established lines of communication or interaction, was also evident in our cases. This was especially clear in the Gulf Coast where, besides the brokered connection between environmentalists and fisherman, the movement diffused rapidly within the established networks that defined the region's separate environmental, commercial, and sport fishing communities.

Relative to our account of these two regional movements, the question of why no movement developed in the Northeast seems entirely straightforward. A big part of the explanation owes to the very different characteristics of the Northeast relative to the other two regions. Although the Gulf Coast and West Coast comprised just three states each, no fewer than seven states were involved in the various proposals slated for the Northeast. The dizzying variety of regulatory and procedural regimes seen across these seven states constituted a formidable institutional impediment to scale shift within the regime. Given these institutional barriers, the presence of a broker/framer who could explain the issue in a way that resonated across so many states becomes all the more important. Unfortunately – at least for the movement – no such figure was forthcoming in the Northeast. The only attempts at brokerage/frame bridging were mounted, not by "neutral" officials, but by activists whose partisan views bred suspicion in potential coalition partners. Coupled with the institutional complexity of the region, the absence of an effective broker/framer probably doomed the movement from the start.

IMPLICATIONS FOR AN UNDERSTANDING OF THE POLICY PROCESS

Although one can find the occasional early work on the topic, the field of policy studies really dates to the 1950s and Harold Lasswell's 1958 call for the development of a systematic science of the policy process. Although work has been continuous since then, the field has really come into its own during the last two decades, with dramatic growth in empirical studies as well as the development of a host of competing theoretical perspectives on the policy process.[1] It is not our intention to review the empirical literature or offer a critical survey of the leading theoretical perspectives. Our aim is much more modest. Although the central focus of our research has been squarely on the twenty communities earmarked for the proposed energy projects, we can also profitably view the broader process we are observing as either the onset of a period of active contention within the U.S. energy policy sector or, more ambitiously, as the emergence and gradual consolidation of several LNG policy subsystems within the broader energy arena. We will have much more to say about the second of these foci, but we begin with a brief discussion of the first.

Energy Policy and the Bush Administration

Given the highly elaborated, increasingly specialized, nature of the energy sector in the United States, it is simply impossible to characterize its workings as governed by anything close to a single integrated policy system. Instead, it would be more accurate to think of energy policy as shaped by a myriad of highly specialized energy policy subsystems (e.g., California

---

[1] For an especially good survey of the leading theories in the field, see Sabatier 2007.

solar energy policy and federal wind power policy). Still, all of these systems are embedded in the same national political and economic environment and thus subject to the same external and internal change pressures that are thought to condition the prospects for stability and change in any area of policy. It is in this more limited sense that we can speak of a broad energy policy sector in the United States.

How should we characterize the workings of this sector? To answer this question, we draw primarily on the Advocacy Coalition Framework (ACF) (Johns 2003; Sabatier and Jenkins-Smith 1988, 1993; Sabatier and Pelkey 1987; Schlager 1995). We begin with a stylized sketch of the ACF, stressing six key elements of the perspective.

- *Policy subsystems.* As we noted in the preceding text, the U.S. energy sector is composed of a very large number of policy subsystems. Policy subsystems are socially constructed arenas composed of actors who seek, on a recurring basis, to influence policy within a fairly narrow substantive domain (e.g., federal policy on offshore oil drilling).
- *Advocacy coalitions.* Each policy subsystem tends to be organized into at least two advocacy coalitions. Advocacy coalitions are aggregations of stakeholders whose policy preferences are loosely aligned, thus constituting a more or less unified voice advocating for specified policy outcomes. These coalitions typically draw adherents from the following groups: elected officials, leaders of interest groups, researchers, agency representatives, and judges.
- *Policy sectors.* Policy sectors are comprised of all the subsystems that together make up a broad policy area (e.g., energy, water, or education). Although the attention of policy actors is normally directed at events within their particular subsystem, during periods of heightened national debate on broad policy questions, we can expect to see advocacy across subsystems.
- *Stability and change.* The prospects for significant change in a given policy subsystem are shaped by the relatively stable features of the external political and economic environments in which it is embedded (e.g., distribution of natural resources or basic governmental structures) and external or internal events that serve to destabilize the subsystem or the broader sector of which it is a part.
- *Punctuated equilibrium.* Because of the relative infrequency of these "destabilizing events," most subsystems tend toward long-term stability, interrupted only rarely by periods of intense contestation,

resulting – even more rarely – in significant policy change. This aspect of the advocacy coalition framework is central to yet another perspective on the policy process, not surprisingly termed "punctuated equilibrium theory" (Baumgartner and Jones 1991, 1993; Dodd 1994; Kingdon 1995; True, Jones, and Baumgartner 2007).

• *Coalition opportunity structures.* Normally the balance of power within a given subsystem affords the advocacy coalitions that comprise it little or no leverage with which to press for significant changes in prevailing policy. The significance of destabilizing events owes to the dramatic impact they have on "coalition opportunity structures." Drawing explicitly on social movement theory, the argument is that the prospects for significant change in policy are linked to the expansion and contraction in political leverage granted the various advocacy coalitions by the force of these destabilizing events.

Armed with this basic framework, we can now use the perspective to describe the significant change in the general energy policy environment that took place during the period of our study. The period opened with the highly contentious and ultimately disputed presidential race between George W. Bush and Al Gore. Lost amidst the general controversy surrounding the contest is the fact that it represented the starkest confrontation between opposing energy visions of any presidential election in history. Gore's tireless efforts to increase awareness of the threat of global warming are well-known. So is his advocacy of a revolution in energy policy, rooted in an embrace of alternative energy sources and a drastic reduction in reliance on gas and oil. It would be hard to imagine a set of energy policy preferences more diametrically opposed to Gore's than those advocated by Bush. As an oilman, Bush brought to the contest a hostile view of the whole climate change debate and a booster's embrace of big oil and gas.

Given these starkly opposing views, Bush's election marked a significant victory, not only for big gas and oil but also for advocates of all manner of traditional energy policies, and a crushing setback for environmentally attuned proponents of energy reform.[2] In short, it represented the kind of destabilizing event that proponents of the ACF stress as necessary to set in

---

[2] Although Bush's immediate predecessor, Clinton, was not as identified with the issue of global warming as was Gore, he was nonetheless a card-carrying liberal regarding energy and environmental policy. Thus in taking office, Bush represented not simply a stark contrast to his opponent in the election, but the man who had shaped federal energy policy for the previous eight years.

motion a period of heightened conflict and significant change in those policy sectors and subsystems impacted by the event. So it proved to be; with one of their kind in the White House, prodevelopment coalitions across a broad range of subsystems pressed for major changes in energy policy. Pressure was applied to open up more offshore sites to oil exploration. Others pushed for a relaxation of environmental restrictions on oil and gas exploration in Alaska. Buoyed by the election, proponents of nuclear power also mobilized, pushing for the first time in a quarter of a century to expand the nation's arsenal of nuclear power plants. Consistent with the ACF, all of the examples cited here were of established policy subsystems destabilized by the 2000 election. We turn now, however, to the role that Bush's election played in catalyzing what we see as the emergence and gradual consolidation of several new policy subsystems that developed in response to the technology of LNG throughout the decade of the 2000s.

Toward a Theory of the Emergence of Policy Subsystems

For all the attention devoted to theorizing the policy process throughout the past two to three decades, all of the major theoretical perspectives in the field betray a singular bias. All are geared to explaining stability and change in established policy subsystems. By contrast, virtually no effort has gone into fashioning an account of the emergence of new subsystems. Because we witnessed this process, we are motivated to close this section with a speculative discussion of the phenomenon as well as a description of the specific regional LNG subsystems born of the events that we have studied.

Perhaps the first thing to say about the phenomenon is that, as social movement analysts, we are struck by just how much the origins of a new policy subsystem resemble the emergence of a social movement. The processes are not simply similar but also often occur in tandem. Take the case of the development of regional movements against LNG that we analyzed in Chapter 5. The emergent movements described in the chapter were simply one component of the broader advocacy coalitions that gradually coalesced in opposition to LNG in the two regions – West Coast and Gulf Coast – throughout the 2000s. Besides traditional movement and interest groups, the emergent coalition included agency officials, media representatives, and select elected officials as well. Developing in parallel with and in opposition to the anti-LNG coalition was a clear pro-LNG coalition, rooted in the gas and oil industry but drawing support as well

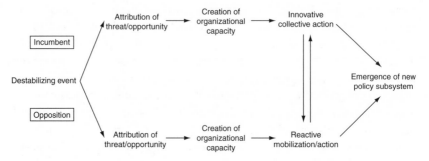

FIGURE 6.2. A dynamic, interactive framework for analyzing the emergence of a policy subsystem

from various media, agency officials, industry researchers, and key elected officials, including those aligned with the Bush administration.

Given the strong family resemblance between these two processes, it makes sense that we draw upon the substantial body of work on movement emergence to fashion a provisional account of the origin of new policy subsystems. More specifically, we turn to a dynamic variant of the political process perspective in order to describe the very similar processes we see reflected in the two phenomena. The perspective is depicted in Figure 6.2.

Figure 6.2 depicts the emergence of a new policy subsystem as the outcome of an unfolding and iterative process of interaction involving at least one "policy innovator" and one actor who mobilizes in opposition to the proposed innovation. In embryonic form, these two actors represent the nucleus of what are expected to develop into the principal advocacy coalitions that are characteristic of all policy subsystems. The figure identifies five linked mechanisms that shape the unfolding process of subsystem emergence. The remainder of this section is given over to a narrative analysis of these five processes as we see them played out in the case of the embryonic LNG subsystems that developed on the West Coast and in the Gulf Coast region.

1. Destabilizing Event

In the case of LNG, we have already identified the destabilizing event that set in motion the process of subsystem emergence: Bush's controversial victory over Gore in the 2000 presidential election. So stark were the differences between the two candidates (and between Bush and his White House predecessor, Bill Clinton) on a range of issues, that actors in a host of policy subsystems attached great significance to the election from the very beginning. In general, the election was expected to either affirm the status quo in most policy subsystems, if Gore prevailed, or introduce powerful change pressures into those same subsystems, should Bush win. The point is that many destabilizing events – such as the 1979 accident at Three Mile Island or the OPEC oil embargo of the mid-1970s, which shattered the policy status quo in the United States with respect to nuclear energy and oil and gas, respectively – come out of the blue, disrupting policy subsystems without warning. The 2000 election and elections in general are not like that. Provided the policy differences between candidates are significant, elections represent a recurring and anticipated threat to policy stability in a host of issue areas.

2. Interpretative Processes and the Collective Attribution of Opportunity/Threat

The ongoing interpretation of events by policy actors shapes the prospects for the emergence of a new subsystem, as it shapes all of social-political life. These continuous processes of sense making and social construction are arguably more important in the case of emergent subsystems, insofar as they require actors to reject institutionalized routines and taken-for-granted assumptions about the world and to fashion new worldviews and lines of action. And yet, for all their importance, these crucial interpretive processes are largely absent from the theories of the policy process. For instance, although, as we have noted, the ACF acknowledges the importance of "destabilizing events," the interpretation of these events is glossed in the framework. That is, the shared understanding that the event in question poses a significant threat to the stability of the policy status quo is presumed to be clear on the face of the event. This assumption strikes us as problematic. While acknowledging some clear exceptions (e.g., the terrorist attacks of 9/11 and the accident at Chernobyl), the threat (or opportunity) posed by a given event is ordinarily not inherent in the event but instead must be constructed by policy actors.

In the case of LNG, it was the oil and gas companies that quickly defined Bush's election as an opportunity to push aggressively for a dramatic expansion of LNG facilities in the United States. Whether this understanding was in place prior to the election or only developed afterward and whether it was a view shared widely in the industry or initially embraced by a single company, we don't know. What we do know is that the formal proposals to build new LNG facilities were forthcoming within a matter of months of Bush's election (Whitmore, Baxter, and Laska 2008). But we are getting ahead of ourselves here. Emerging attributions of opportunity or threat do not, in and of themselves, guarantee innovative collective action. For the latter to occur, one other mechanism is required.

3. Appropriation or Creation of Organizational Capacity

For emerging attributions of threat or opportunity to key innovative policy action, the proponents of the new view must command sufficient organization and numbers to provide a base for mobilization. As a prerequisite for action, would-be policy innovators must either create an organizational vehicle or appropriate an existing one. In short, it isn't enough that some group is motivated to act on the basis of a perceived threat or opportunity that has arisen in its environment, it must also command sufficient organizational capacity to be able to translate that view into action.

In the case of LNG, it was the oil companies that were able to harness their interpretation of the opportunity afforded by the Bush election to a sustained effort to propose and ultimately build dozens of new LNG facilities. Given the speed with which big gas and oil rolled out these proposals in the wake of the 2000 election, it seems clear that whatever in-house opposition there may have been to diverting company resources to LNG, it was overcome quickly and decisively. Thus the shared interpretation of Bush's victory as affording a golden opportunity to promote the development of LNG in the United States came to be harnessed to the organizational resources and political clout of the gas and oil companies.

4. Innovative Collective Action

The result of this combination of perceived opportunities and organizational capacity was the flurry of LNG projects proposed by the oil companies while Bush was in office. Figure 6.3 shows the temporal distribution of the proposals throughout the course of this eight-year span.

5. Reactive Oppositional Mobilization

The final and arguably most important mechanism in the emergence of a new policy subsystem is reactive mobilization by at least one group opposed to the policy innovation embodied in the actions of the initial actor. Absent any such opposition, we lack the basic requirement of competing advocacy coalitions that is characteristic of all policy subsystems.

The dynamics of reactive mobilization bear a bit more attention. Just as the perceived opportunity posed by the initial destabilizing event set in

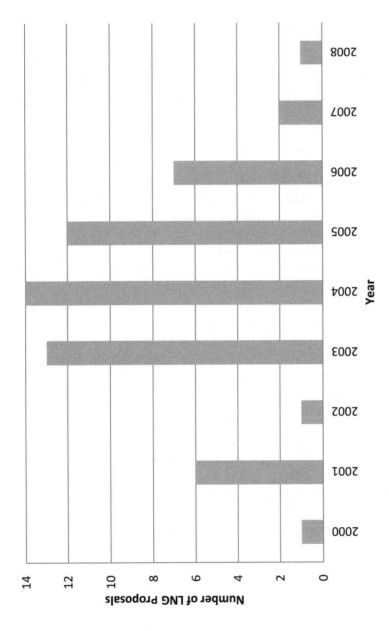

FIGURE 6.3. LNG proposals over time

motion the events that led to the initial policy innovation, it is the perceived threat embodied in that innovation that, in turn, galvanizes reactive mobilization by at least one opposition group. Once these two actors are in place and interacting with each other in a sustained way, the basic requirement and structure of a new policy subsystem is in place. It is very likely, however, that with these two actors in place, other groups or individuals will be motivated to align with either of the two emerging advocacy coalitions. It is through this iterative process that the ultimate structure of the subsystem is gradually consolidated.

This is very much what happened in the case of the two LNG policy subsystems that gradually emerged on the West Coast and Gulf Coast as opposition to the flurry of proposals surfaced, locally at first, but then scaled up to create formidable anti-LNG advocacy coalitions in both regions. As we recounted in great detail in Chapter 5, in the end, both coalitions enjoyed considerable success in blunting the initial momentum achieved by the proponents of LNG. On the Gulf Coast, opposition succeeded in effectively foreclosing one form of LNG (e.g., open loop), while generally embracing the closed-loop alternative. On the West Coast, to date, the opposition coalition has come close to achieving a kind of unofficial moratorium on all LNG projects. More important than the outcome of the conflict in the two regions, however, is the simple fact that, on both coasts, well-defined and stable regional LNG policy subsystems are now in place, where none existed a decade ago.[3]

BACK TO THE FUTURE

Having summarized our answers to the specific questions that structured the research and engaged with the policy process literature, we want to close the book by reengaging the polemic with which we opened it. The research reported here is a product of that broader polemic and its partial expression. We began the book with an extended critique of what we see as the present Ptolemaic state of the field of social movement studies. The very designation of *social movement studies* expresses the narrowness even as it reinforces it. Although the concept of social movement was certainly invoked by pioneering authors such as Tilly, Gamson, Lipsky, and others,

---

[3] With the increase in domestic shale gas production, the LNG policy subsystem is currently being subsumed by a larger natural gas supply policy subsystem. Because of a glut of natural gas in the domestic market, no new LNG plants are being proposed, and those that have been approved and constructed are currently applying for the right to export gas.

none of them saw themselves studying social movements per se. And yet the field has grown so narrow, so movement-centric throughout the years as to distort the original phenomenon of interest – for example, contentious politics – in much the same way that Ptolemy's model of the universe misspecified Earth's position in the cosmos. By selecting movements – and generally large and successful ones at that – for study rather than some other more representative phenomena (e.g., mobilization attempts and populations at risk for a movement), and then generally ignoring the role of other actors in contention, we contend that the field has propounded a view of movements that exaggerates their frequency, characteristic features, and typical impact.

If someone unfamiliar with the field were to immerse him- or herself in the social movement literature, she or he would be forgiven for believing that movements were everywhere in society, large and disruptive in form, and among the central causal forces shaping our world. They would look a great deal like Earth in Ptolemy's model of the universe. We are convinced that the Copernican alternative is much closer to reality. For all the talk of a movement society, most individuals will go through life without ever participating in one. Movements remain a fairly rare phenomenon and, in this sense, confined to the margins of the social-political cosmos rather than positioned at its center. When they do occur, at least in contemporary democracies, they do not tend to look much like our stylized image of 1960s movements, with masses of protestors surging through the streets, damaging property, courting arrest, battling police, and so forth. Most sixties movements did not resemble this stereotype (McAdam et al. 2005). But today's versions betray almost none of the features associated with this stylized image. Consistent with the argument advanced by Meyer and Tarrow (1998), the social movement form has been substantially tamed by being largely institutionalized in the democratic West; ironically, this is a result of the "disruptive" movements of the 1960s and 1970s. Finally, although not unimportant as an influence in social and political life, it is the impact of movements that, we fear, has suffered the most serious distortion as a result of the conventions of social movement scholarship. By focusing primarily on the largest of struggles – civil rights, feminism, and so forth – and then generally ignoring the influence of other simultaneous change processes, movement scholars have "stacked the deck" on the issue of outcomes. Most of the time, we suspect, the impact of popular contention is very modest.

But we do not want our critique of the current narrow state of social movement scholarship to be misunderstood. Reversing Shakespeare's emphasis, we come to praise social movement research, not to bury it. It is just that the best way to do that, in our view, is to restore the field to the Copernican state that prevailed some twenty to thirty years ago. Properly situated in the cosmos, Earth is a fascinating subject for study. So is the social movement; if we truly care about the phenomenon, we will want to broaden out the field of study, take other actors seriously, engage with other relevant literatures (e.g., the policy process), study more representative examples of the form, and in general, put movements in their place by studying episodes of contention rather than the insurgent groups that are but a part of the episode. This is what we have tried to do in the research reported here. We bring the book to a close, then, by highlighting what we see as the key implications of the research. In all, we draw seven lessons from the study.

1. *"It's the [Political] Economy, Stupid"*

At several points in the book we have referenced Jeff Goodwin and Gabriel Hetland's (2010) critical essay on the "strange disappearance of capitalism [or political economy more generally] from social movement studies." As should have been clear from those earlier citations, we are fully in accord with the central thrust of the Goodwin and Hetland critique. Reflecting its Copernican origins, a significant number of early works in the field took the economic roots and dimensions of contentious politics seriously. Today, this emphasis is almost entirely missing from social movement scholarship. The cost of this omission is apparent everywhere in our research. Our cases bear the consistent imprint of the political economy. We will confine ourselves to three examples. For starters, there is the flood of energy proposals – most clearly seen in LNG – that followed in the wake of Bush's election in 2000. As previously noted, this "feeding frenzy" makes it clear that movement actors are not the only ones who mobilize in response to what they see as favorable shifts in political opportunity. Economic actors – in this case gas and oil companies – do as well.

Our recipes predicting nonmobilization and built afford a second clear example of the influence of political economy on the trajectory of our cases. To put it as starkly as possible, company towns (or communities with *similar industry*, as we termed the causal condition) are key to an understanding of which communities do not mobilize and which projects get built. To the extent that a community has come to be economically dependent on some form of energy production, its chances of opposing any new project are greatly reduced and the likelihood of that project coming to fruition are greatly increased.

We close with a final example. Although our results showed that even limited mobilization tended to impact project outcomes negatively, we argued that the distinctive economic calculus underlying the LNG projects helps to account for the surprising result. By proposing far more projects than the market could ever support, the gas and oil companies were looking to jettison those that became economically problematic. Fearing that any popular opposition would occasion delays and cost overruns, the companies were quite willing to abandon those projects without much of a fight. In short, the opposition victories were arguably as much a product of the companies' nonzero, irons-in-the-fire view of the proposals as they were a testament to the strength of opposition forces.

Taken together, these examples support a single, important injunction. If we are ever to understand the dynamics of power, conflict, and change fully, it is absolutely imperative that we build economic actors and their shifting strategic calculus back into our studies of contention. At the risk of running the metaphor into the ground, studying power and contention with little or no reference to economic influences and actors is akin to studying the universe without reference to the sun . . . or at least one of its two most significant suns, the state being the other.

2. Transcending the State

Although economic actors have been perhaps most conspicuous by their absence in studies of contentious politics, other important actors have historically received short shrift as well. Ironically, *the* state is one of those actors. Although the political process model has been justifiably criticized for being overly focused on the state (Armstrong and Bernstein 2008), this does not mean that proponents of the approach have necessarily accorded state actors serious empirical attention in their research. As with most political sociology, empirical accounts of movements often betray a simplistic, reified conception of *the* state, as a unified actor, typically acting in opposition to this or that movement. If political scientists typically ignore or marginalize movements and show real sophistication in decomposing the state into its various component parts – the courts, legislature, state agencies, commissions, and so forth – then social movement scholars reverse the emphasis, offering stylized views of the state while lavishing great analytic attention on various aspects of movements (e.g., tactics, organizational structure, framing, and resources).

The need for a more nuanced understanding of the complex and often contradictory roles played by state actors in episodes of contention can be clearly seen in our results and in our cases. With respect to the former, consider only the results of our fs/QCA analysis for the rejection outcome. The most parsimonious explanation for rejection is a recipe that combines the presence of mobilized opposition (specifically mobilization that is not led by those outside the affected community alone) with significant intergovernmental conflict over the proposed project. This simple combination explains six of our seven "rejected" cases.

Looking beyond the findings, our cases bear the clear imprint of state influence on both sides of the issue. We have already noted the catalytic affect that President Bush had on the issue of LNG. Bush's industry-friendly energy policy stimulated the flood of proposals that set the majority of our cases in motion. Elected officials figured prominently in a host of other cases as well. We have previously noted the importance of Louisiana Governor Blanco's veto of the Main Pass LNG project in helping to "certify" the movement in the region and to set a precedent that made it easier for other governors to follow suit in regard to other proposals.

The especially complicated Crown Landing case turned substantially on a showdown between elected officials in Delaware who opposed the project and those in New Jersey who supported it. Among the key opponents in Delaware were its governor and attorney general, while in New Jersey, a state senator led the charge, when he proposed legislation that would outlaw the use in New Jersey of any credit cards issued by financial institutions based in Delaware. Ultimately, the outcome of the conflict – which turned on a disputed nineteenth-century court case setting the shared boundary of the two states – involved yet another important state actor, the U.S. Supreme Court, which ruled in favor of Delaware's position in the matter.

As important as these rather high-profile actors were to the fate of the aforementioned projects, we were struck throughout this research by the pivotal role played in a good many other cases by lower level, nonelected state officials. By now the key roles played by state and federal fisheries regulators in the Gulf Coast and a CPUC commissioner in brokering and consolidating opposition to LNG in the two regions should be well-known. But these were only the most dramatic and consequential instances of low-level state functionaries significantly shaping the trajectory of this or that case. We could multiply these examples several times over, but we trust the point has been made. Even if *the* state has figured prominently in social movement theory, the *real* state – fragmented, multilevel, and internally contradictory – has rarely informed our empirical understanding of contentious politics.

3. Add Still More Actors

Nor do state and economic actors exhaust the "other" actors whose influence needs to be taken into account in order to understand any given episode of contention. It depends on the case, but in a good many episodes the significant parties to the conflict will include actors beyond the state, economic interests, and emergent grassroots groups. In cases of educational contention, unions are apt to be significant claimants. Churches or other religious institutions are apt to mobilize in response to a wide range of "social issues" (e.g., same sex marriage and abortion). Finally, there is the media. It is a rare episode of contention when the media – or various media – are not centrally involved in shaping its outcome by influencing the public's understanding of and response to the issues and parties in question. This was certainly true in regard to our cases. Although perhaps less influential than the other three actors discussed to this point – grassroots opposition, state officials, and economic interests – the media nonetheless played a key role in several of our cases.

For example, in the case of Cabrillo Port, Malibu reporter Hans Laetz alone wrote more than seventy articles about the project. Laetz broke two stories that played critical roles in turning the tide against the proposal during the summer of 2005: one about supportive comments in the Environmental Impact Statement/Environmental Impact Report falsely attributed to citizens, and another, with the help of investigative work by the Environmental Defense Center, about the Environmental Protection Agency's reversal in regulating the facility's air emissions. Laetz's coverage, originally published in the *Malibu Times*, became so controversial that the newspaper's editor discontinued Laetz's affiliation with the *Malibu Times* during June 2006. However, Laetz quickly found another outlet for his articles in the more liberal *Malibu Surfside News*.

The media played a significant role in several other cases. For instance, in the case of Gulf Landing, Mark Schleifstein, a reporter for *The Times-Picayune* (New Orleans), wrote several influential articles about the potential impact of open-loop regasification systems. These articles helped galvanize opponents of the project. Darryl Malek-Wiley, a well-known environmental activist and Sierra Club member in the New Orleans area, first alerted Schleifstein to the issue. Schleifstein checked with National Marine Fisheries Service personnel about the accuracy of Malek-Wiley's information. Once confirmed, he wrote an article in September 2004 about offshore open-loop LNG proposals and their potential impacts to red drum and snapper populations in the Gulf Coast. After this article, *The Times-Picayune*, fishing groups, and environmental organizations all took negative stances on open-loop systems and encouraged the governor to veto these proposals. Moreover, Schleifstein's reporting was picked up by reporters at the *Mobile Press-Register* in Alabama, who alerted Mobile Baykeeper about the potential impacts posed by Compass Port's open-loop regasification technology. These cases highlight the critical importance played by media coverage in alerting the general public to issues and mobilizing opposition.

We want to close this section on a more general note. It should be clear that our point here is not about the media or any particular actor. It is a more general plea that studies of contention take account of all relevant parties to the conflict. It is incumbent on the analyst to know his or her case well enough to identify the relevant actors and find ways to assess their influence on the emergence, development, and ultimate resolution of the case in question.

4. Putting Movements in Their Place

Among the many fictions supported by the analytic conventions of the field is the fact that most movements somehow take place everywhere and nowhere in particular. The field's preoccupation with "national" struggles may be largely responsible for this fiction. Although it is true that environmental activism, lesbian, gay, bisexual, and transgender activism, or contentious claims making about any other broad issue does occur in many places, it is never placeless. Instead, regardless of the geographic reach of the underlying issue, contention is always embedded in a local context. As we noted in Chapter 5, the notion of a national movement is something of a fiction. Almost always, national movements are nothing more than coalitions of local struggles. This was certainly the case with the African American civil rights movement of the 1950s and 1960s. The familiar roll call of iconic campaigns is little more than a list of local movements: Montgomery, Little Rock, Albany, Birmingham, and Selma. However, attention to place is rare in contemporary movement studies. This inattention to place is yet another reflection of the increasing insularity of much social movement scholarship. How so? Because it persists in the face of a growing body of work by geographers that seeks to understand contention, precisely in relation to issues of space and place (Glassman 2002; Harvey 2000; Miller 2000, 2004; Shepard and Hayduk 2002). But with the field embedded primarily in sociology – at least in the United States – most movement scholars aren't even aware of the work of these geographers. It is time, however, that movement analysts, in whatever discipline, take the need to put movements in their place seriously in an effort to take account of the multiple ways in which the local context powerfully shapes the prospects for and the forms and outcomes of contention. In choosing to study twenty discrete communities, this was among our central theoretical and empirical goals.

The results of our research, we think, speak to the wisdom of this decision. The powerful imprint of place is reflected everywhere in our findings. We use two of our findings to illustrate the pervasiveness of this influence. The first of these was touched on in the preceding text. We are speaking of the singularly strong influence of *similar industry* on the likelihood of opposition mobilization and the probability that the proposed project would get built. In discussing these findings in the preceding text, we emphasized the influence of political economy. But these results also speak to a deeper, more amorphous way in which place shapes the prospects for local contention. Company towns are not simply a reflection of economic interests and influence. They are places whose way of life, its rhythm and pace, collective memories, and shared sense of identity bear the self-serving imprint of the dominant employer in the community. It is no wonder then that projects proposed for communities with similar industries are less likely to engender opposition and more likely to get built in the end.

For much the same reason, we find that communities that have previously mobilized against a similar land-use proposal are, not surprisingly, very likely to do so again, resulting ultimately in the rejection or withdrawal of a project. Once again, we interpret these findings as a reflection of the way that place – as a repository for shared experiences and memories – shapes the prospects for and outcomes of contention.

5. Fieldwork, Fieldwork, Fieldwork

Throughout the years, a number of movement scholars have urged analysts to invest more deeply in ethnography and other fieldwork methods (e.g., interview and archival work) (Auyero 2003; Binder 2002; McAdam, McCarthy, and Zald 1988; Mische 2008). Based on the value of our experiences in the field, we renew that call here. Only by doing so can we hope to move beyond statistical relationships in order to understand social movements and contention in the deeper Weberian sense of *verstehen*. This is more or less where we see the field at present. That is, throughout the past twenty-five to thirty years, we have amassed an impressive body of empirical work that has, on several key theoretical issues, yielded fairly consistent statistical relationships between static variables. So, for example, numerous analysts have shown that prior social ties predict movement recruitment (Briet, Klandermans, and Kroon 1987; Broadbent 1986; Diani 1995; Fernandez and McAdam 1988; Gould 1993, 1995; Klandermans and Oegema 1987; McAdam 1986; McAdam and Paulsen 1993; Snow, Zurcher, and Ekland-Olson 1980). Others have shown that movements tend to develop, not under conditions of isolation and social disorganization as predicted by traditional collective behavior theory, but where established grassroots organizations and informal networks afford aggrieved populations the mobilizing structures that facilitate emergent collective action (Evans 1979; Gould 1995; McAdam 1999 [1982]; Morris 1984; Zhao 1998).

These are impressive lines of work, attesting to serious and sustained inquiry by a generation of movement scholars. Together they represent a body of knowledge about social movements that simply did not exist in, for example, 1970. The central thrust of this empirical work contradicts the received theoretical wisdom regarding movements circa 1970. For all the virtues of this work, however, the nature of the understanding it affords us of movement dynamics is still fairly limited. Writing about the network studies of movement recruitment, Passy (2003: 22) captures the limitation as follows: "we are now aware that social ties are important for collective action, but we still need to theorize ... the actual role of networks." That is, the statistical studies of social movements that have tended to dominate the field historically have helped us understand the when and where of contention, but not the why or how. For that we will need to invest in more qualitative, fieldwork-based methods. The good news, as described in some detail in Chapter 2, is that by fully exploiting online data sources, researchers can learn a tremendous amount about sites and cases before heading off into the field. By practicing what we wound up calling "advanced preparation fieldwork," we were able to reduce the time spent in any one community dramatically while still coming to know our cases in the ethnographic sense of the term. For anyone who has shied away from traditional fieldwork because of the time and expense they thought was involved in the method, you just might have to come up with another excuse for not investing in the technique.

We close with one final, perhaps obvious but nonetheless important, point. In the previous section, we urged movement scholars to take seriously the myriad ways that place shapes contention. Needless to say, this aim can only be realized through the simultaneous embrace of fieldwork methods. We are convinced that the general ignorance of place in the study of contention owes in large part to the relative paucity of ethnographic fieldwork by movement scholars.

6. Mechanisms and the Dynamics of Contention

We can afford to be very brief in stating the sixth of our prescriptions for the field. For our sixth prescription is almost more an extension of the fifth than a separate injunction. In the previous section, we noted that the largely statistical body of knowledge produced to date is pretty good on the when and where of contention, but much less useful when it comes to the how and why. To correct this imbalance, we have argued for a deeper investment in the full range of fieldwork methods. But to what, pray tell, should field-workers attend? Those specific social processes, or mechanisms, that shape the emergence, development, and resolution of concrete episodes of contention.

Reflecting the field's traditional preoccupation with quantitative methods, much of the discussion and debate stimulated by McAdam, Tarrow, and Tilly's 2001 call for a "mechanistic" turn in the study of contention has focused on measurement issues (Earl 2008; Falleti and Lynch 2008; Lichbach 2008; McAdam, Tarrow, and Tilly 2001; Staggenborg 2008). This has led, in some cases, to a narrow insistence that absent quantitative measures, the study of mechanisms is futile. Although we would never discourage quantitative approaches to the problem, the narrow equation of mechanisms with quantifiable measures, is misguided and ignorant of forms of explanation other than those that rely on inferential statistics (Brady and Collier 2010; Collier and Elam 2008; Freedman 2006, 2010; George and Bennett 2005; Tarrow 2010). Even if they never invoked the term, anthropologists and historians have been interrogating mechanisms for generations. One hardly needs a precise quantitative measure of a given mechanism to divine its presence or discern its impact. This is where the thick description of fieldwork comes into play. Consider only our attribution of the important causal role played by brokerage in our narrative accounts of the emergence of regional movements against LNG in the Gulf Coast and West Coast. Our evidence for invoking the mechanism came from numerous informants across a number of cases in both regions. These informants included the self-reports of several of the brokers as well as corroborating accounts by a number of the activists whose participation in the emerging regional movement resulted from the connections forged by the broker. It is frankly hard to imagine how a narrow, quantitative measure of brokerage could improve upon the triangulated narrative accounts of the presence and impact of the mechanism that we were able to amass through our fieldwork.

7. Redefining the Phenomenon of Interest

Finally and, in our view, most importantly, we hope that our research strategy has demonstrated the great value to be derived from abandoning the study of those exceedingly rare events we have called *social movements* in favor of a comparative analysis of mobilization attempts, episodes of contention, or, as in our case, communities at risk for mobilization. It is not that the case study of single movements has produced nothing of value. On the contrary, many of the most celebrated works in the field are works of this sort. There is no question that we can learn a great deal about social movements and the broader dynamics of contention from the case-study approach. The problem comes when we want to generalize from these clearly rare, atypical events to broader phenomenon. Does it really make sense to imagine that we can understand the dynamics of emergent collective action only from those exceedingly rare instances that produce large amounts of it and sustain it over long stretches of time? Isn't it more likely that our understanding of emergent mobilization will benefit from looking at a broader array of cases, ranging from the successful movement at one end of the continuum to instances of resolute inaction on the other end?

Our basic point would seem to apply with even more force when it comes to a consideration of movement outcomes. How can we ever hope to characterize the social change potential inherent in the movement form accurately by only studying the largest, most sustained, and by implication, most consequential, of struggles? As we have noted at several points in this book, one of the real surprises to emerge from this research is the influence that even low levels of opposition appear to exert on the ultimate fate of the projects. But because this finding came from a comparative study of twenty at-risk communities rather than a single case study of a large, sustained struggle, we are that much more inclined to interpret the results as telling us something generalizable about movements rather than about an artifact of nonrandom case selection.

The value of taking nonevents as seriously as events is perhaps best seen in our efforts – reported in Chapter 3 – to derive causal recipes for nonmobilization. For scholars interested in the dynamics of emergent contention, why wouldn't it be as important to understand the kind of factors and dynamic processes that inhibit mobilization, as those that appear to facilitate it? If I were an organizer, I might actually be more concerned with the former than the latter. As Alinsky (1989) reminded us long ago, mobilizing people – especially those burdened by disadvantage and feelings of fatalism – is so hard that you simply can't afford initial setbacks. Confronting these challenges, one wishes to avoid failure at all cost. In such a situation, knowing which factors predict nonmobilization could be very valuable. But it is the *scholarly* benefits that have motivated this research and the expansive reorientation of the field we are calling for here. Fortunately, in recent years a handful of scholars have begun to take the dynamics of negative cases seriously, resulting in at least the beginnings of a literature on the topic (Auyero and Swistun 2009; Bell 2010; Sherman 2011).

We are convinced that none of the prescriptions on offer here would have a more salutary effect on the study of contention than a fundamental redefinition of the basic phenomenon of interest. Whether one conceives of this phenomenon as mobilization attempts, communities at risk for mobilization, or episodes of contention matters less than the resolve to broaden the field by moving movements from the center of our scholarly cosmos to its margins.

We close with a final perhaps irreverent thought. The study of social movements emerged as a vibrant new area of scholarly inquiry by looking outward and engaging with a broad range of other fields and disciplines, including history, organizational studies, political science, and the study of political economy. The initial breadth and outward-looking focus of the enterprise made it an exciting new interdisciplinary field of study. Thirty years later, the field looks very different, at least to our eyes. It is infinitely bigger than it was in the late 1970s and early 1980s. But its size, as we noted in Chapter 1, has created its own problems. The field is now sufficiently large as to support two specialty journals, several book series, and countless specialized social movement conferences each year. In the abstract, this dramatic expansion of a field that, for all intents and purposes, did not exist thirty years ago should be cause for celebration. In practice, however, what all this growth has encouraged is more insularity and less engagement with the very areas of inquiry that helped shape the field at the outset. The field is now sufficiently large as to function essentially as its own audience. This makes it increasingly hard for insights from other areas or disciplines to find their way into the field or for challenges to the conceptual or methodological conventions of the field to be mounted from within. Having been socialized into those conventions early in one's career, it is harder to see them for what they are: scholarly norms that obscure as much as they reveal. For any field to remain vital, these norms must be constantly open to challenge and informed by insights from other areas of study. As the field has grown and the tendency toward insularity has become more pronounced, the possibilities for this kind of intellectual revitalization have dimmed.

In the face of these developments, what is one to do? Thomas Jefferson was famously quoted as saying, "God forbid we should ever be twenty years without ... a rebellion." He was speaking about political systems, not academic fields. But as the tendencies toward narrow academic special-ization grow stronger every day, we find ourselves in sympathy with the Jeffersonian impulse. This book and the research reported in its pages are motivated by this impulse.

Appendix A
Additional Community Data Collected Not Used in Causal and Outcome Condition Scoring

Percent female
Median age
Percent households with children
Percent completed high school
Percent non-English speakers at home
Poverty levels
Home value
League of Conservation Voters scores for elected representatives (U.S. Congress and Senate)
Environmental nonprofits per capita
Health nonprofits per capita
Environmental nonprofit revenues
Health nonprofit revenues
Political party affiliation
Newspaper presidential endorsements (2000, 2004)
Percent employment in extractive industries
Regional location

Appendix B
Raw Data and Methods for Scoring Causal Conditions

Generally scores were assigned to quantitative variables corresponding to their percentile value among the twenty communities. Typically, in fuzzy set/Qualitative Comparative Analysis one must designate a point of maximum ambiguity. In this method, the median or 50th percentile is designated as the point of maximum ambiguity. Cases fell in a set for a specific variable if the value of the variable was greater than 0.5. Under this condition the community received a score of 0.6. If the variable value was higher than the 80th percentile, it received a score of 0.8. If the variable value was above the 99th percentile, it received a score of 1. Cases fell out of the set with values of 0.4 if the value was above the 40th percentile and below the 50th percentile, 0.2 if the value was above the 20th percentile and below the 40th percentile, and 0 if the value was below the 20th percentile. Scores were cross-checked against natural break points in the data and typically matched these quite nicely. Some variables (noted in the following text) were assigned scores limited to set membership (1) or nonmembership (0), which reflected if a certain condition was present.

After individual variables were assigned fuzzy scores, they were often combined to form the conceptual causal condition. For example, civic and organizational capacity was constructed using the fuzzy scores for values associated with a community's general education level, voter turnout rates, and number of nonprofits per capita. Most commonly, we combined these by adding the fuzzy scores assigned to individual variables and then reassigning fuzzy scores to the sum according to the procedure detailed in the preceding text in which scores corresponded to percentile rank.

MOBILIZATION

Community	Letters (score)	Speakers (score)	Coordinated Appearances (score)	Public Meetings (score)	Protest (score)	Lawsuit (score)	Intensity Sum (score)	Collective Behavior?	Noninstitutional Activity?	Final Score
Aiken County, SC	12 (0.6)	21 (0.6)	0 (0)	1 (0.2)	2 (0.6)	1	3 (0.6)	Y	Y	0.6
Brazoria County, TX	5 (0.2)	7 (0.2)	1 (0.6)	0 (0)	0 (0)	0	1 (0.2)	Y	N	0.2
Cameron Parish, LA (Sabine Pass)	0 (0)	1 (0)	0 (0)	0 (0)	0 (0)	0	0 (0)	N	N	0
Cameron Parish, LA (Gulf Landing)	0 (0)	1 (0)	2 (0.6)	0 (0)	2 (0.6)	1	2.2 (0.6)	Y	Y	0.6
Cameron Parish, LA (Creole Trail)	0 (0)	1 (0)	0 (0)	0 (0)	0 (0)	0	0 (0)	N	N	0
Cassia County, ID	5 (0.2)	5 (0.2)	1 (0.6)	1 (0.2)	0 (0)	0	1.2 (0.2)	Y	N	0.2
Claiborne County, MS	2 (0.2)	9 (0.4)	1 (0.6)	2 (0.8)	1 (0.6)	1	3.6 (0.8)	Y	Y	0.8
Essex County, MA	19 (0.6)	26 (0.6)	0 (0)	1 (0.2)	0 (0)	0	1.4 (0.2)	Y	N	0.2

Community	Letters (score)	Speakers (score)	Coordinated Appearances (score)	Public Meetings (score)	Protest (score)	Lawsuit (score)	Intensity Sum (score)	Collective Behavior?	Noninstitutional Activity?	Final Score
Gloucester County, NJ	6 (0.6)	13 (0.4)	0 (0)	1 (0.2)	0 (0)	0	1.2 (0.2)	Y	N	0.2
Long Beach, CA	65 (0.8)	22 (0.6)	5 (0.8)	1 (0.2)	2 (0.6)	0	3 (0.6)	Y	Y	0.6
Malibu, CA	123 (0.8)	138 (1)	6 (1)	3 (0.8)	5 (0.8)	0	4.2 (0.8)	Y	Y	0.8
Mobile County, AL	47 (0.6)	38 (0.6)	0 (0)	1 (0.2)	1 (0.6)	0	2 (0.4)	Y	Y	.6
New Castle County, DE	9 (0.6)	20 (0.6)	1 (0.6)	1 (0.2)	1 (0.6)	0	2.6 (0.6)	Y	Y	.6
Providence, RI	15 (0.6)	17 (0.6)	0 (0)	1 (0.2)	0 (0)	0	1.4 (0.2)	Y	N	.2
Riverside County, CA	5 (0.2)	72 (0.8)	5 (0.8)	1 (0.2)	1 (0.6)	0	2.6 (0.6)	Y	Y	.6
San Patricio County, TX (Cheniere)	0 (0)	0 (0)	0 (0)	0 (0)	0 (0)	0	0 (0)	N	N	0
San Patricio, TX (Vista del Sol)	2 (0.2)	0 (0)	0 (0)	0 (0)	0 (0)	0	0 (0.2)	N	N	0
Solano County, CA	192 (0.8)	120 (0.8)	1 (0.6)	3 (0.8)	4 (0.8)	0	3.8 (0.8)	Y	Y	.8
Ventura County, CA	415 (1)	138 (1)	5 (0.8)	3 (0.8)	8 (1)	0	4.4 (1)	Y	Y	1
Whatcom County, WA	2 (0.2)	4 (0.2)	0 (0)	0 (0)	0 (0)	0	0.4 (0)	N	N	0

SAFETY THREAT

To score this condition we used two variables. First, we used the proximity of the project to residential populations. Proximity matters because accidents at these sites are thought to affect most severely those who are nearby through fires caused by explosion, radiation, floods caused by dams breaking, and so forth. For this reason, projects are designed to minimize the chances that accidents will affect populations outside the site. We also used the population density of the area surrounding the project. This represents the fact that, in order to minimize risk, projects are often placed in remote locations. Several project opponents argued that this should be the case. Scores for these two variables were assigned to communities relative to other communities in the study based on percentage rank as described in the preceding text. They were then added together, and this final sum was given a final score representative of this final percentage rank.

Community	Proximity (miles)	Proximity Score	Population Density (population/ sq. mile)	Density Score	Sum of Proximity and Density Scores	Final Score
Aiken County, SC	12	0.4	133	0.4	0.8	0.2
Brazoria County, TX	0	1	174	0.4	1.4	0.6
Cameron Parish, LA (Sabine Pass)	15	0.2	8	0	0.2	0
Cameron Parish, LA (Gulf Landing)	38	0	8	0	0	0
Cameron Parish, LA (Creole Trail)	2	0.6	8	0	0.6	0
Cassia County, ID	6	0.4	8	0	0.4	0
Claiborne County, MS	6	0.4	24	0.2	0.6	0
Essex County, MA	13	0.4	1,442	0.8	1.2	0.6
Gloucester County, NJ	1	1	798	0.6	1.6	0.6
Long Beach, CA	2	0.6	9,966	1	1.6	0.6
Malibu, CA	14	0.2	629	0.6	0.8	0.2
Mobile County, AL	11	0.4	327	0.6	1	0.6
New Castle County, DE	2	0.6	1,007	0.8	1.4	0.6
Providence, RI	1	1	986	0.8	1.8	1
Riverside County, CA	0	1	239	0.6	1.6	0.6
San Patricio County, TX (Cheniere)	2	0.6	97	0.2	.8	0.2
San Patricio, TX (Vista del Sol)	2	0.6	97	0.2	0.8	0.2
Solano County, CA	1	1	479	0.6	1.6	0.6
Ventura County, CA	14	0.2	425	0.6	0.8	0.2
Whatcom County, WA	2	0.6	79	0.2	.8	0.2

Notes: Proximity is the distance of the project in miles from the closest residence. In the case of offshore projects, it is the distance in miles from the nearest coastline. Population density was calculated by dividing the community's land area in square miles by the population.

PROPERTY THREAT

To score this condition the first step was to determine whether the project was close to any residential property. Thus we assigned fuzzy scores to proximity as in the safety condition. We assume concern about property threats is relatively unfounded in cases in which projects are so far away they are neither visible nor likely to generate any day-to-day nuisance. Thus distant communities were assigned scores less than 0.5, making them out of the set, while nearby communities were assigned scores greater than 0.5, putting them in the set. To further differentiate between communities, we used data on the median home value. To assign a score to home value, we normalized the value of the median single-family detached home by dividing it by the median income. This provides us with a sense of the degree of investment in property. We then assigned a final fuzzy score based upon the percentile rank of the home value. Communities in close proximity were scored as follows: a value score of 0.8 or 1 received a final score of 1, a value is 0.4 or 0.6 received a final score of at 0.8, and a value score of 0.2 or 0 received a final score of 0.6. For communities out of the set, a value score of 0.8 or 1 was assigned a 0.4, a values score of 0.6 or 0.4 was assigned a final score of 0.2, and a value score of 0.2 or 0 was assigned a final score of 0.

Community	Home Value	Median Income	Home Value/Median Income	Value Score	Proximity (miles)	Proximity Score	Final Score
Aiken County, SC	87,600	$37,889	2.31	0.2	12	0.4	0
Brazoria County, TX	88,500	$48,632	1.82	0	0	1	0.6
Cameron Parish, LA (Sabine Pass)	59,600	$34,232	1.74	0	15	0.2	0
Cameron Parish, LA (Gulf Landing)	59,600	$34,232	1.74	0	38	0	0
Cameron Parish, LA (Creole Trail)	59,600	$34,232	1.74	0	2	0.6	0.6
Cassia County, ID	83,100	$33,322	2.49	0.4	6	0.4	0.2
Claiborne County, MS	48,200	$22,615	2.13	0.2	6	0.4	0
Essex County, MA	362,131	$57,280	6.32	0.8	13	0.4	0.4
Gloucester County, NJ	155,323	$57,214	2.71	0.4	1	1	0.8
Long Beach, CA	274,903	$38,975	7.05	0.8	2	0.6	0.8
Malibu, CA	1,000,000	$102,031	9.80	1	14	0.2	1
Mobile County, AL	94,170	$34,000	2.77	0.6	11	0.4	0.4
New Castle County, DE	172,252	$55,344	3.11	0.6	2	0.6	0.2
Providence, RI	165,458	$45,634	3.63	0.6	1	1	0.8
Riverside County, CA	223,924	$45,135	4.96	0.6	0	1	0.8
San Patricio County, TX (Cheniere)	66,000	$34,836	1.89	0.2	2	0.6	0.6
San Patricio, TX (Vista del Sol)	66,000	$34,836	1.89	0.2	2	0.6	0.6
Solano County, CA	266,992	$60,847	4.39	0.6	1	1	0.8
Ventura County, CA	487,961	$65,260	7.48	0.8	14	0.2	0.4
Whatcom County, WA	155,700	$40,005	3.89	0.6	2	0.6	0.8

Sources: 2000 U.S. Census and American Communities Survey.

THREAT BY PROJECT TYPE

The literature on risk provides guidance about how communities may perceive threat based on project type. From this literature, we assign fuzzy scores to show the relative levels of perceived threat as follows: nuclear storage and related industries receives a 1, nuclear power receives a 0.8, onshore liquefied natural gas (LNG) projects and oil refineries receive a score of 0.6, offshore LNG projects receive a 0.4, hydroelectric receives a 0.2, and wind receives a score of 0. Sixteen communities are in the set of those facing a threatening project based on project type alone.

Community	Type	Final Score
Aiken County, SC	Nuclear Fuel Refabrication	1
Brazoria County, TX	LNG Terminal	0.6
Cameron Parish, LA (Sabine Pass)	LNG Terminal	0.6
Cameron Parish, LA (Gulf Landing)	LNG Terminal (offshore open-loop)	0.6
Cameron Parish, LA (Creole Trail)	LNG Terminal	0.6
Cassia County, ID	Wind Power Generation	0
Claiborne County, MS	Nuclear Power	0.8
Essex County, MA	LNG Terminal (offshore)	0.4
Gloucester County, NJ	LNG Terminal	0.6
Long Beach, CA	LNG Terminal	0.6
Malibu, CA	LNG Terminal (offshore)	0.4
Mobile County, AL	LNG Terminal (offshore open-loop)	0.6
New Castle County, DE	LNG Terminal	0.6
Providence, RI	LNG Terminal	0.6
Riverside County, CA	Hydroelectric	0.2
San Patricio County, TX (Cheniere)	LNG Terminal	0.6
San Patricio, TX (Vista del Sol)	LNG Terminal	0.6
Solano County, CA	LNG Terminal	0.6
Ventura County, CA	LNG Terminal (offshore)	0.4
Whatcom County, WA	Oil Refinery/Electricity Cogeneration Project	0.6

RISK

We arrive at a final score for whether a proposal poses significant risk or
threat to the community by taking the average of the fuzzy-set scores for
each component of risk. We take the average to indicate the overall
strength of various arguments that could be made against a project, though
not all communities draw on each argument. Overall, eight communities
were in the set of those facing a risky project.

Community	Safety Threat	Property Threat	Project Type	Sum of Safety, Property, and Type	Final Score
Aiken County, SC	0.2	0	1	1.2	0.2
Brazoria County, TX	0.6	0.6	0.6	1.8	0.6
Cameron Parish, LA (Sabine Pass)	0	0	0.6	0.6	0
Cameron Parish, LA (Gulf Landing)	0	0	0.6	0.6	0
Cameron Parish, LA (Creole Trail)	0	0.6	0.6	1.2	0.2
Cassia County, ID	0	0.2	0	0.2	0
Claiborne County, MS	0	0	0.8	0.8	0
Essex County, MA	0.6	0.4	0.4	1.4	0.4
Gloucester County, NJ	0.6	0.8	0.6	2	0.6
Long Beach, CA	0.6	1	0.6	2.2	0.8
Malibu, CA	0.2	0.4	0.4	1	0.2
Mobile County, AL	0.6	0.2	0.6	1.4	0.4
New Castle County, DE	0.6	0.8	0.6	2	0.6
Providence, RI	1	0.8	0.6	2.4	1
Riverside County, CA	0.6	0.8	0.2	1.6	0.6
San Patricio County, TX (Cheniere)	0.2	0.6	0.6	1.4	0.4
San Patricio, TX (Vista del Sol)	0.2	0.6	0.6	1.4	0.4
Solano County, CA	0.6	0.8	0.6	2	0.6
Ventura County, CA	0.2	0.4	0.4	1	0.2
Whatcom County, WA	0.2	0.8	0.6	1.6	0.6

CIVIC AND ORGANIZATIONAL CAPACITY

This condition consists of three variables – nonprofits per one thousand residents, percent college educated, and voter turnout. Raw statistics are in the following text. These values are converted to fuzzy-set scores by assigning them relative to other communities according to their percentile value. The final score is calculated by adding the individual fuzzy-set scores together, finding a new percentile value, and assigning the score accordingly.

Community	Nonprofits/1,000	Nonprofits Score	Education	Education Score	Voter Turnout	Voter Turnout Score	Sum of Scores	Final Score
Aiken County, SC	1.77	0.2	19.9	0.4	65	0.6	1.2	0.4
Brazoria County, TX	1.83	0.4	19.6	0.4	55	0	0.8	0.2
Cameron Parish, LA (Sabine Pass)	0.70	0	7.9	0	59	0.2	0.2	0
Cameron Parish, LA (Gulf Landing)	0.70	0	7.9	0	59	0.2	0.2	0
Cameron Parish, LA (Creole Trail)	0.70	0	7.9	0	62	0.4	0.4	0.2
Cassia County, ID	1.87	0.6	13.9	0.2	72	0.6	1.4	0.4
Claiborne County, MS	1.86	0.6	18.9	0.2	54	0	0.8	0.2
Essex County, MA	2.64	0.6	37.6	0.8	73	0.6	2	0.6
Gloucester County, NJ	1.44	0	21.7	0.6	74	0.8	1.4	0.4
Long Beach, CA	2.31	0.6	24.8	0.6	59	0.2	1.4	0.4
Malibu, CA	9.54	1	59.4	1	60	0.2	2.2	0.8
Mobile County, AL	2.16	0.6	21.5	0.6	71	0.6	1.8	0.6
New Castle County, DE	5.49	0.8	32.7	0.8	66	0.6	2.2	0.8
Providence, RI	5.69	0.8	26.7	0.6	61	0.4	1.8	0.6
Riverside County, CA	1.71	0.2	18.8	0.2	73	0.6	1	0.2
San Patricio County, TX (Cheniere)	1.67	0.2	13.0	0	42	0	0.2	0
San Patricio County, TX (Vista del Sol)	1.67	0.2	13.0	0	42	0	0.2	0
Solano County, CA	1.82	0.4	25.7	0.6	69	0.6	1.6	0.6
Ventura County, CA	2.25	0.6	29.2	0.8	79	1	2.4	1
Whatcom County, WA	3.14	0.8	27.2	0.6	76	0.8	2.2	0.8

Sources: Nonprofits per 1,000 people was calculated using counts of nonprofit organizations in the 2000 Business Master File and dividing by the community's population in 2000. Education represents the percentage of individuals aged 25 and older with a college degree as reported by the 2000 U.S. Census or the American Communities Survey. Voter turnout is the percentage of registered voters who voted in the most recent presidential election prior to the announcement.

POLITICAL OPPORTUNITY

A community's membership in the set of those with ample political opportunity was assessed relative to other communities. Political opportunity is thought to rest on three elements. First, political opportunity is most directly present when decision makers are elected officials. This provides citizens a chance to pressure decision makers through the electoral process. Thus the first measure of political opportunity is the proportion of the decision-making body comprised of elected officials. Based on a percentile-ranking method, we assigned a score for this proportion. Second, whether those elected officials are up for reelection affects the amount of political pressure that can be exacted. During a campaign, citizens can pressure officials to take a firm position on a siting issue, making it a campaign issue rather than just one of many issues that a representative faces. Political opportunity should be enhanced if a reelection campaign is present in the community, if so this received a score of 1 and received a 0 otherwise. Finally, the jurisdiction of the elected officials matters. Through the electoral process, citizens have more power over local officials because they make up a larger proportion of the electorate. The jurisdiction score reflects this fact. A community receives a 1 if the elected officials were local, 0.6 if there was at least one local elected official, and 0 if there were no local elected officials on the decision-making body.

Community	Proportion Elected Officials	Proportion Score	Reelection	Jurisdiction	Sum	Final Score
Aiken County, SC	0	0	0	0	0	0
Brazoria County, TX	0.69	0.8	0	1	1.8	0.6
Cameron Parish, LA (Sabine Pass)	0	0	0	0	0	0
Cameron Parish, LA (Gulf Landing)	0.5	0.6	0	0.6	1.2	0.6
Cameron Parish, LA (Creole Trail)	0	0	0	0	0	0
Cassia County, ID	0	0	0	0	0	0
Claiborne County, MS	0	0	0	0	0	0
Essex County, MA	0.5	0.6	0	0.6	1.2	0.6
Gloucester County, NJ	0	0	0	0	0	0
Long Beach, CA	0.5	0.6	1	1	2.6	0.8
Malibu, CA	0.6	0.6	0	0.6	1.2	0.6
Mobile County, AL	0.67	0.8	1	0.6	2.4	0.8
New Castle County, DE	0	0	0	0.6	0.6	0.4
Providence, RI	0	0	0	0	0	0
Riverside County, CA	0.31	0.6	1	1	2.6	0.8
San Patricio County, TX (Cheniere)	0	0	0	0	0	0
San Patricio County, TX (Vista del Sol)	0	0	0	0	0	0
Solano County, CA	1	1	1	1	3	1
Ventura County, CA	0.6	0.6	0	0.6	1.2	0.6
Whatcom County, WA	0.14	0.6	1	0.6	2.2	0.6

Source: These data come from author's expert knowledge of the case generated through newspaper analysis and field interviews.

HARDSHIP

This variable is scored as described in the preceding text. Scores were assigned individually to the unemployment figures for each community and the income levels of each community. Unemployment is merely the percentage of the civilian labor force that is unemployed. Income scores were assigned relative to membership in the set of low-income communities. Thus the lowest-income communities were fully in the set scoring a value of 1, and the highest-income communities were fully out of the set scoring a value of 0. These scores were then summed together, and this sum served as the basis for assignment of the final score based on percentile value of the sum.

Community	Unemployment	Unemployment Score	Income	Income Score	Sum of Scores (unemployment + income)	Final Score
Aiken County, SC	5.9	0.4	$37,889	0.6	1	0.2
Brazoria County, TX	5.4	0.2	$48,632	0.4	0.6	0
Cameron Parish, LA (Sabine Pass)	4.6	0	$34,232	1	1	0.2
Cameron Parish, LA (Gulf Landing)	4.6	0	$34,232	1	1	0.2
Cameron Parish, LA (Creole Trail)	4.6	0	$34,232	1	1	0.2
Cassia County, ID	5.2	0.2	$33,322	1	1.2	0.6
Claiborne County, MS	18	1	$22,615	1	2	1
Essex County, MA	5.6	0.2	$57,280	0.2	0.4	0
Gloucester County, NJ	7	0.6	$57,214	0.4	1	0.2
Long Beach, CA	8	0.8	$38,975	0.6	1.4	0.6
Malibu, CA	2.8	0	$102,031	0	0	0
Mobile County, AL	9.8	0.8	$34,000	1	1.8	0.8
New Castle County, DE	6.1	0.6	$55,344	0.4	1	0.2
Providence, RI	6.1	0.6	$45,634	0.4	1	0.2
Riverside County, CA	8.3	0.8	$45,135	0.4	1.2	0.6
San Patricio County, TX (Cheniere)	7.2	0.6	$34,836	0.8	1.4	0.6
San Patricio County, TX (Vista del Sol)	7.2	0.6	$34,836	0.8	1.4	0.6
Solano County, CA	5	0.2	$60,847	0.2	0.4	0
Ventura County, CA	5.7	0.4	$65,260	0.2	0.6	0
Whatcom County, WA	7.4	0.6	$40,005	0.4	1	0.2

Sources: Unemployment is the percentage of civilians aged 16 and older in the labor force who are currently unemployed as reported by the 2000 U.S. Census or the American Communities Survey. Income is the median household income for the community as reported by the U.S. Census or the American Communities Survey.

REGIONAL SATURATION AND IRONS IN THE FIRE

These causal conditions are scored based on data collected about the state of the market at the time of approval of each project. To measure regional saturation, we looked at the number of projects approved and under construction at the time of the project's approval in the project's region (East Coast, West Coast, or Gulf Coast) by another company since the company's announcement of the project. Specifically, a case received a score of 1 if three projects were approved and under construction in the region by other companies since the announcement of the proposal, a score of 0.67 if two projects were approved and under construction, a score of 0.33 if only one project was approved and under construction, and a score of 0 if no projects were approved and under construction. To capture irons in the fire, we measured the number of projects by the same company approved in the region at the time of the given project's approval since the company's announcement of the project. Specifically, a case received a score of 1 if more than one other project by the same company had been approved in the region since the company's announcement of the project, a score of 0.4 if only one other project by the same company had been approved, and a score of 0 if no other projects by the same company had been approved in the region.

Community	Proposal Announcement	Proposal Approval or Withdrawal	Projects by Other Companies	Saturation Score	Projects by Same Company	Irons Score
Brazoria County, TX	September 2001	June 2004	None	0	None	0
Cameron Parish, LA (Sabine Pass)	May 2003	December 2004	Gulf Gateway	0.33	Freeport	0.4
Cameron Parish, LA (Gulf Landing)	November 2003	February 2005	Gulf Gateway Freeport Sabine Pass	1	None	0
Cameron Parish, LA (Creole Trail)	January 2005	May 2006	Gulf Gateway Freeport Sabine Pass	1	Freeport Sabine Pass Corpus Christi	1
Essex County, MA	June 2004	February 2007	None	0	None	0
Gloucester County, NJ	December 2003	June 2006	Northeast Gateway Neptune	0.67	None	0
Mobile County, AL	March 2004	June 2006	Gulf Gateway Freeport Sabine Pass	1	Golden Pass	0.4
San Patricio County, TX (Cheniere)	May 2003	April 2005	Gulf Gateway	0.33	Freeport Sabine Pass	0.67
San Patricio County, TX (Vista del Sol)	October 2003	June 2005	Gulf Gateway Freeport Sabine Pass	1	None	0
Solano	May 2002	January 2003	None	0	None	0

INTERGOVERNMENTAL CONFLICT AND OUTSIDE LEADERSHIP

These causal conditions are scored based on data collected about the opposition effort using the Final Environmental Impact Statement (EIS) and interviews. We measured intergovernmental conflict as follows: a case received a score of 1 if a lawsuit was filed by state or federal officials related to the project, 0.8 if interagency disputes were referred to the White House Council on Environmental Quality and a state or federal agency denied a permit associated with the project, 0.6 if state or federal agencies wrote letters critical of the proposal during the EIS review process, and 0 otherwise. We scored outside leadership dichotomously: a project received a score of 1 if opposition was led by outsiders absent much local support and a 0 otherwise.

Case	Evidence of Intergovernmental Conflict	Intergovernmental Conflict	Evidence of Outside Leadership	Outside Leadership (absent local)
Aiken	None	0	Opposition led by state and national antinuclear groups	1
Brazoria	None	0	None	0
Cameron Creole Trail	None	0	None	0
Cameron Gulf Landing	Dispute between lead agency and National Marine Fisheries Service referred to White House Council on Environmental Quality	0.8	Opposition led by state and regional environmental and fishing organizations	1
Cameron Sabine Pass	None	0	None	0
Cassia	None	0	None	0
Claiborne	None	0	Opposition led by national antinuclear groups	1
Essex	Letters from National Marine Fisheries Services and state agencies regarding potential effects on fisheries	0.6	None	0
Gloucester	Dispute over jurisdictional authority between Delaware and New Jersey to court	1	Opposition led by Delaware environmental groups and state agencies	1
Long Beach	Dispute over jurisdictional authority between state and federal agency to court	1	None	0

Malibu	Letters from state agencies regarding potential air-pollution impacts	0.6	None	0
Mobile	Letters from governor's office regarding potential effects on fisheries	0.6	None	0
New Castle	Dispute over jurisdictional authority between Delaware and New Jersey to court	1	None	0
Providence	Letters from attorney general and other state agencies regarding potential safety concerns	0.6	None	0
Riverside	Denial of permit by State Water Board	0.8	None	0
San Patricio Cheniere	None	0	None	0
San Patricio Vista del Sol	None	0	None	0
Ventura	Letters from state agencies regarding potential air-pollution impacts	0.6	None	0
Whatcom	None	0	None	0

DATA SOURCES

Community	Start Year	Demographic Data[a]	Demographic Data Year	Election Year[b]	Newspaper(s)
Aiken County, SC	2001	Census	2000	2000	*Augusta Chronicle* (Augusta, GA)
Brazoria County, TX	2001	Census	2000	2000	*Brazosport Facts* (Clute, TX)
Cameron Parish, LA (Sabine Pass)	2003	Census	2000	2000	*Cameron Parish Pilot*
Cameron Parish, LA (Gulf Landing)	2003	Census	2000	2000	*Cameron Parish Pilot*
Cameron Parish, LA (Creole Trail)	2004	Census	2000	2004	*Cameron Parish Pilot*
Cassia County, ID	2002	Census	2000	2000	*Times-News* (Twin Falls, ID)
Claiborne County, MS	2003	ACS	2002	2000	*Vicksburg Post* (Vicksburg, MS)
Essex County, MA	2002	Census	2000	2000	*Gloucester Daily Times*
Gloucester County, NJ	2005	ACS	2004	2004	*News Journal* (Wilmington, DE)
Long Beach, CA	2004	ACS	2003	2004	*Press-Telegram*
Malibu, CA	2004	Census	2000	2004	*Los Angeles Times* (Los Angeles, CA)
Mobile County, AL	2004	ACS	2003	2004	*Press-Register* (Mobile, AL)
New Castle County, DE	2004	ACS	2003	2004	*News Journal* (Wilmington, DE)

(cont.)

Community	Start Year	Demographic Data[a]	Demographic Data Year	Election Year[b]	Newspaper(s)
Providence, RI	2004	ACS	2002	2004	*Providence Journal*
Riverside County, CA	2003	ACS	2003	2000	*Press-Enterprise*
San Patricio County, TX (Cheniere)	2004	ACS	2003	2004	*Corpus Christi Caller-Times*
San Patricio County, TX (Vista del Sol)	2003	Census	2000	2000	*Corpus Christi Caller-Times*
Solano County, CA	2003	Census	2000	2000	*San Francisco Chronicle* (San Francisco, CA)
Ventura County, CA	2004	ACS	2004	2000	*Ventura County Star*
Whatcom County, WA	2002	Census	2000	2000	*Bellingham Herald* (Bellingham, WA)

[a] Demographic data includes data for population, median income, unemployment, percent with a college degree, median home value, home ownership (ratio of owner-occupied housing relative to all occupied housing). Census indicates the U.S. Decennial Census conducted by the U.S. Bureau of the Census. ACS indicates the American Communities Survey periodically conducted by the U.S. Bureau of the Census, which provides more current estimates on a subset of communities.

[b] Election data was secured from a variety of sources. County clerks are required to collect and report such data to the states. Therefore this figure was often taken directly from county clerk or state Web sites.

Appendix C
Interview Sources by Case

Unless otherwise noted, interviews were conducted by second author.

Date	Interviewee	Affiliation
Aiken County, SC[a]		
July 21, 2009	Tim Harris	Nuclear Regulatory Commission
August 4, 2009	Mal McKibben	Citizens for Nuclear Technology Awareness
August 4, 2009	Clint Wolfe	Citizens for Nuclear Technology Awareness
August 4, 2009	Walt Joseph	Savannah River Site Heritage Foundation
August 4, 2009	Tom Clements	Friends of the Earth
August 5, 2009	John Clark	South Carolina Energy Office
Brazoria County, TX		
May 20, 2009	Phyllis Saathoff	Port Freeport
May 20, 2009	Alaina Olinde	Quintana Island resident
May 20, 2009	Tonya Barker	City of Quintana
May 21, 2009	Tobey Davenport	Port Freeport
May 21, 2009	Sharron Stewart	Environmental activist
May 22, 2009	Teresa Cornelison	Quintana Island resident
May 22, 2009	Michael Johns	Freeport
May 22, 2009	Wallace Neeley	City of Quintana Mayor

(*cont.*)

Date	Interviewee	Affiliation
Cabrillo Port (Malibu, Ventura County, CA)		
April 25, 2007	n/a	State official
May 2, 2007	Kathi Hann	BHP
May 3, 2007	Owen Bailey	Sierra Club Great Coastal Places
July 17, 2007	Tony Skinner	Tri-Counties Building and Construction Trades
July 17, 2007	Steven Weiner	Tri-Counties Building and Construction Trades
November 4, 2007	Anne Sheehan	State Lands Commission
November 20, 2007	Susan Jordan	California Coastal Protection Network
January 25, 2008	n/a	State official
February 29, 2008	n/a	State official
April 3, 2008	Renee Klimscak	BHP
March 20, 2007	Andy Stern	Malibu council member
May 4, 2007	Hans Laetz	*Malibu Surfside News*
July 18, 2007	Jolene Dodson	Assistant to the Brosnans
July 18, 2007	Lynn Griffin	Malibu Volunteer Coordinator for Cabrillo Port opposition
May 2, 2007	John Flynn	Ventura County Supervisor
May 2, 2007	Thomas Holden	Oxnard mayor
May 2, 2007	Gloria Roman	Saviers Road Design Team
May 2, 2007	William Terry	Saviers Road Design Team
May 3, 2007	Denis O'Leary	Oxnard School District
May 3, 2007	Nancy Lindholm	Oxnard Chamber of Commerce
May 30, 2007	Maricela Morales	Mayor, Port Hueneme/Central Coast Alliance United for a Sustainable Economy

(cont.)

Date	Interviewee	Affiliation
July 17, 2007	Shirley Godwin	Saviers Road Design Team
July 17, 2007	Charles Godwin	Saviers Road Design Team
July 17, 2007	Don Facciano	Ventura County Taxpayers Association

Cameron Parish, LA (Gulf Landing)

December 10, 2008	Leslie March	Sierra Club's Delta Chapter
December 17, 2008	Alex Williamson	Tulane Environmental Law Center
January 15, 2009	Charlie Smith	Louisiana Charter Boat Association
January 16, 2009	Aaron Viles	Gulf Restoration Network
January 16, 2009	Mark Schleifstein	New Orleans *The Times-Picayune*
January 22, 2009	Greg Koehler	Gulf Landing

Cameron Parish, LA (Sabine Pass, Creole Trail)

January 9, 2009	Patricia Outtrim	Cheniere Energy
January 12, 2009	Carl Griffith	Former Jefferson County judge (TX)
January 12, 2009	George Swift	Southwest Louisiana Economic Development Alliance
January 13, 2009	Adam McBride	Lake Charles Terminal and Harbor District
January 14, 2009	Scott Trahan	Cameron Parish Police Juror
January 14, 2009	Ernest Broussard	Cameron Parish Planning and Development

Cassia County, ID[b]

August 20, 2009	Chip Thompson	*Twin Falls Times News*
August 24, 2009	Roald Doskeland	Windland, Inc.
August 24, 2009	Michael Heckler	Windland, Inc.
August 24, 2009	Bob Masterson	Idaho Power

(*cont.*)

Date	Interviewee	Affiliation
August 25, 2009	David Parrish	Idaho Fish and Game
August 26, 2009	Jim Wahlgren	Committee Against Windmills in Albion
August 26, 2009	Don Bowden	Former Albion city council member
August 27, 2009	Clay Handy	Cassia County Commissioner
August 27, 2009	Wendy Reynolds	BLM Burley Field Office

Claiborne County, MS

April 1, 2009	Eric Oesterle	Nuclear Regulatory Commission
April 3, 2009	Paul Gunter	Nuclear Information and Resource Service
April 6, 2009	Greg Sparks	Entergy Nuclear
April 6, 2009	Ken Hughey	Entergy Nuclear
April 6, 2009	Mike Bowling	Entergy Nuclear
April 6, 2009	Tom Williamson	Entergy Nuclear
April 7, 2009	James Johnston	Claiborne County Administrator
April 7, 2009	A.C. Garner	NAACP Claiborne County Chapter
April 7, 2009	Charlie Mitchell	*Vicksburg Post*
April 7, 2009	Danny Barrett, Jr.	*Vicksburg Post*
April 7, 2009	Emma Crisler	*Port Gibson Reveille*
April 7, 2009	Kenneth Ross	Port Gibson Main Street, Inc.
April 14, 2009	Tami Dozier	Nuclear Regulatory Commission

Crown Landing

September 20, 2006	Robert Kopka	Federal Energy Regulatory Commission
August 5, 2009	John Burzichelli	Mayor, Township of Paulsboro and New Jersey State Legislator

(*cont.*)

Date	Interviewee	Affiliation
January 10, 2010	Paul Fader	Chief Counsel, New Jersey Governor
August 3, 2009	Alan Muller	Green Delaware
August 3, 2009	Phil Cherry	Delaware Department of Natural Resources and Environmental Control
August 3, 2009	Jeff Montgomery	*Wilmington News Journal*
August 4, 2009	Christine Whitehead	New Castle Civic League
August 4, 2009	Dave Bailey	New Castle Civic League
August 4, 2009	Al Denio	Sierra Club, Delaware Chapter
August 4, 2009	Kenneth Kristl	Widener Law School

Essex County, MA

July 29, 2009	Bob Bradford	North Shore Chamber of Commerce
July 29, 2009	John Bell	Former Mayor, city of Gloucester
July 29, 2009	Angela Sanfilippo	Gloucester Fishermen's Wives Association
July 30, 2009	Rob Bryngelson	Excelerate Energy

Long Beach, CA

November 3, 2008	Chris Garner	City of Long Beach Gas and Oil Department
November 3, 2008	Tom Modica	City of Long Beach
November 4, 2008	Val Lerch	City of Long Beach
November 4, 2008	Bry Myown	Long Beach Citizens for Utility Reform
November 5, 2008	Robert Kanter	Port of Long Beach
November 5, 2008	Richard Steinke	Port of Long Beach
November 5, 2008	Stacey Crouch	Port of Long Beach
November 6, 2008	Richard Slawson	Los Angeles/Orange Counties Building and Construction Trades

(*cont.*)

Date	Interviewee	Affiliation
Mobile County, AL		
March 24, 2009	Steve Lawless	ConocoPhillips
April 8, 2009	Jeff Rester	Gulf States Marine Fisheries Commission
April 9, 2009	Edwin Lamberth	Coastal Conservation Association
April 9, 2009	Ginny Russell	Mobile Area Chamber of Commerce
April 9, 2009	Casi Callaway	Mobile Baykeeper
April 10, 2009	George Crozier	Dauphin Island Sea Lab
April 10, 2009	Steve Heath	Retired Alabama Department of Conservation
April 10, 2009	Jeff Collier	Mayor, town of Dauphin Island
May 4, 2009	Ben Raines	*Mobile Press Register*
Providence County, RI[c]		
September 20, 2006	Dave Swearingen	Federal Energy Regulatory Commission
June 22, 2009	John Torgan	Save the Bay
June 23, 2009	Christopher D'Ovidio	Conservation Law Foundation
June 23, 2009	Raymond Gallison	State Representative
June 24, 2009	Paul Roberti	Attorney general's office
June 24, 2009	David Manning	National Grid/KeySpan
Riverside County, CA		
February 4, 2009	Doug Pinnow	Candidate, Elsinore Valley Municipal Water District
February 5, 2009	John Lloyd	Elsinore Valley Municipal Water District
February 5, 2009	Greg Morrison	Elsinore Valley Municipal Water District
February 5, 2009	Rexford Wait	Nevada Hydro

(cont.)

Date	Interviewee	Affiliation
February 6, 2009	Robert Magee	Mayor, city of Lake Elsinore
February 6, 2009	Autumn DeWoody	Inland Empire Waterkeeper

San Patricio County, TX (Cheniere, Vista del Sol)

May 18, 2009	Mark Scott	Corpus Christi City Council member
May 18, 2009	Johnny French	Retired U.S. Fish and Wildlife Service
May 18, 2009	Terry Simpson	San Patricio County Judge
May 19, 2009	John LaRue	Port of Corpus Christi
May 19, 2009	Frank Brogan	Port of Corpus Christi
May 19, 2009	Bill Green	Port of Corpus Christi
May 19, 2009	Georgia Neblett	Former mayor, Port Aransas
May 15, 2009	Paul Friedman	Federal Energy Regulatory Commission
July 6, 2009	De McLallen	Community Relations Consultant to ExxonMobil

Solano County, CA

August 1, 2006	Don Parker	Vallejo Fire Chief
August 3, 2006	Tony Intintoli	Vallejo Mayor
August 3, 2006	Vicki Gray	Vallejo Community Planned Renewal
August 3, 2006	Matthias Gafni	*Vallejo Times-Herald*
August 3, 2006	Chris Denina	*Vallejo Times-Herald*
August 3, 2006	Jack Bungart	*Vallejo Times-Herald*
October 5, 2006	Stephanie Gomes	Vallejo Community Planned Renewal
October 6, 2006	Rod Boschee	Vallejo Community Planned Renewal
October 16, 2006	Linda Engelman	Vallejo Planning Commission

(*cont.*)

Date	Interviewee	Affiliation
October 19, 2006	Elena DuCharme	Vallejo Community Planned Renewal
November 2, 2006	Liz Burleigh	Bystander
November 2, 2006	Bryant Burleigh	Bystander
March 22, 2007	Connie Klimisch	Vallejo Chamber of Commerce
March 22, 2007	Kurt Henke	Firefighters' Union
May 11, 2007	Alan Mawdsley	Bechtel
October 2, 2007	Jack Fugett	Bechtel consultant

Whatcom County, WA[d]

July 21, 2009	Fred Felleman	Ocean Advocates
July 21, 2009	Alen Fiksdal	Energy Facility Site Evaluation Council
July 22, 2009	Michael Abendhoff	BP Cherry Point
July 22, 2009	Doug Erickson	State Representative
July 23, 2009	Elie Friloeb	Birch Bay resident
July 23, 2009	Dan McShane	Whatcom County Council
July 23, 2009	Wendy Steffensen	RE Sources

Regional Movements

May 19, 2006	Monica Schwebs	California Energy Commission
May 23, 2006	Keith Lesnick	Maritime Administration
September 20, 2006	Alysa Lykens	Federal Energy Regulatory Commission
September 22, 2006	Bob Corbin	Department of Energy
November 29, 2006	Mark Prescott	U.S. Coast Guard
November 27, 2007	Pat Mason	California Foundation on the Environment and the Economy

(*cont.*)

Date	Interviewee	Affiliation
April 3, 2008	Richard Myers	California Public Utilities Commission
January 13, 2009	Oliver G. "Rick" Richard, III	Former Federal Energy Regulatory Commission Commissioner
April 15, 2009	Miles Croom	National Marine Fisheries Service
July 16, 2009	Carla Garcia Zendejas	Mexican environmental attorney
July 23, 2009	Carl Weimer[e]	Pipeline Safety Trust
August 11, 2009	Rory Cox	Pacific Environment
January 6, 2010	Robert Godfrey	Save Passamaquoddy Bay
February 11, 2010	Adrienne Esposito	Citizens Campaign for the Environment
February 12, 2010	Suedeen Kelly	Former Federal Energy Regulatory Commission Commissioner

[a] All interviews for this case were conducted by R. Gong.
[b] All interviews for this case were conducted by the first author.
[c] All interviews for this case were conducted by R. Wright.
[d] All interviews for this case were conducted by R. Wright.
[e] Interview conducted by R. Wright.

Bibliography

Adam, Barry D. 1987. *The Rise of a Gay and Lesbian Movement*. Boston: Twayne.

Adeola, Frances O. 2000. "Cross-National Environmental Injustice and Human Rights Issues: A Review of Evidence in the Developing World." *American Behavioral Scientist* 43: 686–706.

Aldrich, Daniel P. 2008. *Site Fights: Divisive Facilities and Civil Society in Japan and the West*. Ithaca, NY: Cornell University Press.

Alinsky, Saul. 1989. *Rules for Radicals: A Practical Primer for Realistic Radicals*. New York: Vintage Books.

Amenta, Edwin. 2006. *When Movements Matter: The Townsend Plan and the Rise of Social Security*. Princeton, NJ: Princeton University Press.

Amenta, Edwin, Sheera Joy Olasky, and Neal Carren. 2005. "Age for Leisure? Political Mediation and the Impact of the Pension Movement on U.S. Old-Age Policy." *American Sociological Review* 70: 516–39.

Anderson-Sherman, Arnold and Doug McAdam. 1982. "American Black Insurgency and the World Economy: A Political Process Model." Pp. 165–88 in *Ascent and Decline in the World System*, ed. Edward Friedman. Beverly Hills, CA: Sage Publications.

Andrews, Kenneth T. 1997. "The Impacts of Social Movements on the Political Process: A Study of the Civil Rights Movement and Black Electoral Politics in Mississippi." *American Sociological Review* 62: 800–19.

2001. "Social Movements and Policy Implementation: The Mississippi Civil Rights Movement and the War on Poverty, 1965–71." *American Sociological Review* 66: 71–95.

2004. *Freedom Is a Constant Struggle*. Chicago: University of Chicago Press.

Andrews, Kenneth T. and Michael Briggs. 2006. "The Dynamics of Protest Diffusion: Movement Organization, Social Networks and News Media in the 1960 Sit-ins." *American Sociological Review* 71 (5): 752–7.

Antosh, N. 2003. "LNG Deal Advances to Solid Footing – First such Terminal in U.S. in 20 Years." *Houston Chronicle*, December 23.

Arjomand, Said. 1988. *The Turban for the Crown: The Islamic Revolution in Iran.* New York: Oxford University Press.

Armstrong, Elizabeth A. 2005. "From Struggle to Settlement: The Crystallization of a Field of Lesbian/Gay Organizations in San Francisco, 1969–1973." Pp. 161–188 in *Social Movements and Organization Theory*, ed. Gerald F. Davis, Doug McAdam, W. Richard Scott, and Mayer N. Zald. New York: Cambridge University Press.

Armstrong, Elizabeth A. and Mary Bernstein. 2008. "Culture, Power, and Institutions: A Multi-Institutional Politics Approach to Social Movements." *Sociological Theory* 26: 74–99.

Auyero, Javier. 2003. *Contentious Lives: Two Argentine Women, Two Protests, and the Quest for Recognition.* Durham, NC and London: Duke University Press.

Auyero, Javier and Débora Alejandra Swistun. 2009. *Flammable.* New York: Oxford University Press.

Baker, M. 2002. "Port Approves Deal for Natural Gas Facility." *Brazosport Facts*, December 14.

Baldassare, Delia and Mario Diani. 2007. "The Integrative Power of Networks." *American Journal of Sociology* 113: 735–80.

Banaszak, Lee Ann. 1996. *Why Movements Succeed or Fail.* Princeton, NJ: Princeton University Press.

Baumgartner, Frank R. and Bryan D. Jones. 1991. "Agenda Dynamics and Policy Subsystems." *Journal of Politics* 53: 1044–74.

1993. *Agendas and Instability in American Politics.* Chicago: University of Chicago Press.

Beissinger, Mark. 2001. *Nationalist Mobilization and the Collapse of the Soviet State: A Tidal Approach to the Study of Nationalism.* Cambridge: Cambridge University Press.

Bell, Shannon Elizabeth. 2010. "Fighting King Coal: The Barriers to Grassroots Environmental Justice Movement Participation in Central Appalachia." PhD diss., Department of Sociology, University of Oregon. Eugene.

Bennett, Andrew. 2010. "Process Tracing and Causal Inference." Pp. 207–19 in *Rethinking Social Inquiry*, ed. Henry E. Brady and David Collier. Lanham, MD: Rowman and Littlefield.

Beshur, A. 2005. "Cheniere LNG Terminal Gets OK: Firm to Build near Gregory." *Corpus Christi Caller-Times*, April 14.

Binder, Amy J. 2002. *Contentious Curricula.* Princeton, NJ: Princeton University Press.

Blumer, Herbert. 1939. "Collective Behavior." Pp. 219–88 in *Principles of Sociology*, ed. Robert E. Park. New York: Barnes and Noble.

Boholm, A., ed. 2004. *Facility Siting: Risk, Power and Identity in Land Use Planning.* Toronto: Earthscan Canada.

Bohstedt, John and Dale Williams. 1987. "The Diffusion of Riots: The Patterns of 1766, 1795 and 1801 in Devonshire." *Journal of Interdisciplinary History* 19: 1–24.

Boli, John and George M. Thomas. 1997. "World Culture in the World Polity: A Century of International Non-Governmental Organizations." *American Sociological Review* 62: 171–90.

Border Power Plants Working Group. 2004. "Global LNG Summit Flyer," http://www.borderpowerplants.org/Final%20Global%20LNG%Summit%20invitation.pdf (accessed August 18, 2008).

———. 2008. "Global LNG Summit Flyer 2004," http://www.borderpowerplants.org/Final%20Global%20LNG%20Summit%20invitation.pdf (accessed August 18, 2008).

———. 2009. "Border Plants Working Group 2009," http://www.borderpowerplants.org/ (accessed December 16, 2009).

Boudet, Hilary S. 2010. "Contentious Politics in Liquefied Natural Gas Facility Siting." PhD diss., Interdisciplinary Program in Environmental Research, Stanford University, Stanford, CA.

Boudet, Hilary S. and L. Ortolano. 2010. "A Tale of Two Sitings: Contentious Politics in Liquefied Natural Gas Facility Siting in California." *Journal of Planning Education and Research* 30: 5–21.

Brady, Henry E. and David Collier. 2004. *Rethinking Social Inquiry*. Lanham, MD: Rowman and Littlefield.

———. 2010. *Rethinking Social Inquiry*. 2nd ed. Lanham, MD: Rowman and Littlefield.

Briet, Martien, Bert Klandermans, and Frederike Kroon. 1987. "How Women Become Involved in the Women's Movement in the Netherlands." Pp. 44–63 in *The Women's Movement of the United States and Western Europe: Consciousness, Political Opportunities and Public Policy*, ed. Mary Fainsod Katzenstein and Carol McClurg Mueller. Philadelphia: Temple University Press.

Broadbent, Jeffrey. 1986. "The Ties that Bind: Social Fabric and the Mobilization of Environmental Movements in Japan." *International Journal of Mass Emergencies and Disasters* 4: 227–53.

Buechler, Steven M. 1990. *Women's Movements in the United States*. New Brunswick, NJ: Rutgers University Press.

Bunce, Valerie. 1999. *Subversive Institutions: The Design and the Destruction of Socialism and the State*. Cambridge: Cambridge University Press.

Burstein, Paul. 1993. "Explaining State Action and the Expansion of Civil Rights: The Civil Rights Act of 1964." *Research in Political Sociology* 6: 117–37.

———. 1998. "Bringing the Public Back In: Should Sociologists Consider the Impact of Public Opinion on Public Policy?" *Social Forces* 77: 27–62.

California Energy Commission. 2005. "Safety Advisory Report on the Proposed Sound Energy Solutions LNG Terminal at the Port of Long Beach, California." Sacramento, CA.

California State Lands Commission. 2007. "Cabrillo Port Final EIR Hearing." April 9.

Carruthers, D. V. 2007. "Environmental Justice and the Politics of Energy on the U.S.-Mexico Border. *Environmental Politics* 16: 394–413.

Casselman, B. 2009. "Sierra Club's Pro-Gas Dilemma: National Group's Stance Angers On-the-Ground Environmentalists in Several States." *Wall Street Journal*, December 31.

Chriss, C. 2005a. *General Development of Malibu, 1970 to 1991*, http://www.malibucomplete.com/mc_history_dev_1970s-91_general.php (accessed August 11, 2008).

2005b. *Malibu vs. Coastal Commission*, http://www.malibucomplete.com/mc_history_malibucity_coastal.php (accessed August 11, 2008).

Christensen N. 2009. "Oregon Legislature: Bill Could Stop LNJ Projects." *The Hillsboro Argus*, January 13.

Citizens for Nuclear Technology Awareness. 2010. "Citizens for Nuclear Technology Awareness 2010," http://www.c-n-t-a.com/home.htm (accessed December 11, 2010).

Clemens, Elisabeth S. 1997. *The People's Lobby: Organizational Innovation and the Rise of Interest Group Politics in the United States, 1890–1925.* Chicago: University of Chicago Press.

Collier, David and Colin Elam. 2008. "Qualitative and Multi-Method Research: Organizations, Publication, and Reflections on Integration." Pp. 779–95 in *The Oxford Handbook of Political Methodology*, ed. Janet Box-Steffensmeier, Henry E. Brady, and David Collier. Oxford: Oxford University Press.

Conell, Carol and Samuel Cohn. 1995. "Learning from Other People's Actions: Environmental Variation and Diffusion in French Coal Mining Strikes, 1890–1935." *American Journal of Sociology* 101: 366–403.

Conservation Law Foundation. 2010. "LNG Terminal Siting: Developing a Regional Strategy," http://www.clf.org/work/CECC/LNGterminalsiting/index.html (accessed August 1, 2010).

Coppens, Tom. 2011. "Understanding Land-Use Conflicts in Strategic Urban Projects: Lessons from Gent Sint-Pieters." Pp. 189–211 in *Strategic Spatial Projects: Catalysts for Change*, ed. Stijn Oosterlynck, Jef Van den Broeck, Louis Albrechts, Frank Moulaert, and Ann Verhetsel. New York: Routledge.

Costain, Anne W. 1992. *Inviting Women's Rebellion: A Political Process Interpretation of the Women's Movement.* Baltimore, MD: Johns Hopkins University Press.

Cox, R. 2005. *Voices of Opposition Along the LNG Fast-Track.* Pacific Environment, www.pacificenvironment.org/article.php?id=142 (accessed March 16, 2008).

Cox, R. and R. Freehling. 2008. "Collision Course: How Imported LNG Will Undermine Clean Energy in California," http://www.pacificenvironment.org/article.php?id=2710 (accessed March 16, 2008).

Cress, Daniel M. and David A. Snow. 2000. "The Outcomes of Homeless Mobilization: The Influence of Organization, Disruption, Political Mediation and Framing." *American Journal of Sociology* 105: 1063–1104.

Crisler, E. 2004. "Grand Gulf." *Port Gibson Reveille*, March 11.

2009. "Editor's Comments: Entergy Is Plus." *Port Gibson Reveille*, April 2.

Crosby, E. 2005. *A Little Taste of Freedom: The Black Freedom Struggle in Claiborne County, Mississippi.* Chapel Hill: University of North Carolina Press.

Dahl, Robert A. 1967. *Pluralist Democracy in the United States.* Chicago: Rand McNally.

Dalton, Russell, Alix Van Sickle, and Steven Weldon. 2009. "The Individual-Institutional Nexus of Protest Behavior." *British Journal of Political Science* 40: 51–73.

Davies, Scott and Linda Quirke. 2005. "Providing for the Priceless Student: Ideologies of Choice in an Emerging Educational Market." *American Journal of Education* 111: 523–47.

Davis, Gerald F. and Doug McAdam. 2000. "Corporations, Classes, and Social Movements." Pp. 195–238 in *Research in Organizational Behavior*, vol. 22, ed. Barry Straw and Robert I. Sutton. Oxford: Elsevier Science.

Davis, Gerald F., Doug McAdam, W. Richard Scott, and Mayer N. Zald, eds. 2005. *Social Movements and Organization Theory*. New York: Cambridge University Press.

Dear, M. 1992. "Understanding and Overcoming the NIMBY Syndrome." *Journal of the American Planning Association* **58** (3): 288–300.

della Porta, Donatella. 1995. *Political Movements, Political Violence and the State*. New York: Cambridge University Press.

Diani, Mario. 1995. *Green Networks: A Structural Analysis of the Italian Environmental Movement*. Edinburgh: Edinburgh University Press.

Dodd, Lawrence C. 1994. "Political Learning and Political Change: Understanding Development across Time." Pp. 331–64 in *The Dynamics of American Politics*, ed. Lawrence C. Dodd and Calvin Jillson. Boulder, CO: Westview Press.

Domhoff, G. William. 1970. *The Higher Circles*. New York: Random House.

Dosh, Paul. 2009. "Tactical Innovation, Democratic Governance, and Mixed Motives: Popular Movement Resilience in Peru and Ecuador." *Latin American Politics and Society* **51**: 87–118.

Downs, Anthony. 1957. *An Economic Theory of Democracy*. New York: Harper and Row.

Drezner, Daniel W. 1999. *The Sanctions Paradox: Economic Statecraft and International Relations*. Cambridge: Cambridge University Press.

Earl, Jennifer. 2000. "Methods, Movements, and Outcomes: Methodological Difficulties in the Study of Extra-Movement Outcomes." *Research in Social Movements, Conflicts and Change* **22**: 3–25.

    2008. "An Admirable Call to Improve, but Not Fundamentally Change Our Collective Methodological Practices." *Qualitative Sociology* **31**: 355–60.

Eccleston, C. H. 2008. *NEPA and Environmental Planning: Tools, Techniques, and Approaches for Practitioners*. Boca Raton, FL: CRC Press.

Edelman, Lauren B., Gwendolyn Leachman, and Doug McAdam. 2010. "On Law, Organizations and Social Movements." *Annual Review of Law and Social Science* **6**: 653–85.

Eisinger, Peter. 1973. "The Conditions of Protest Behavior in American Cities." *American Political Science Review* **67**: 11–28.

Environmental Defense Center. 2002. "Environmental Defense Center Case Docket: 25th Anniversary Edition, 1977–2002." http://www.edcnet.org/docket.pdf (accessed August 11, 2008).

Environmental Protection Agency. 2007. "National Environmental Policy Act (NEPA): Basic Information." http://www.epa.gov/compliance/basics/nepa.html (accessed July 10, 2007).

Evans, Sara. 1979. *Personal Politics*. New York: Knopf.

Falleti, Tulia G. and Julia Lynch. 2008. "From Process to Mechanism: Varieties of Disaggregation." *Qualitative Sociology* **31**: 333–40.

Federal Energy Regulatory Commission. 2003. Draft EIS Hearing on Freeport LNG Project, December 9.

    2004a. Draft EIS Hearing on Sabine Pass LNG and Pipeline Project. September 21.

    2004b. Draft EIS for the Cheniere Corpus Christi LNG Project. Washington, DC.

    2004c. Draft EIS KeySpan LNG Facility Upgrade Project. Washington, DC.

    2004d. Meeting Record Vista del Sol LNG Terminal and Pipeline Project. Washington, DC.

    2005. Long Beach LNG Import Project Draft EIS/EIR and Draft Port Master Plan Amendments No. 20. Washington, DC.

Fernandez, Roberto and Doug McAdam. 1988. "Social Networks and Social Movements: Multiorganizational Fields and Recruitment to Mississippi Freedom Summer." *Sociological Forum* 3: 357–82.

Ferree, Myra Marx and Frederick D. Miller. 1985. "Mobilization and Meaning: Toward an Integration of Social Movement Theory." *Sociological Inquiry* 55: 38–51.

Feuer, Lewis. 1969. *The Conflict of Generations*. New York: Basic Books.

Finkel, Steven E., Edward N. Muller, and Karl Dieter-Opp. 1989. "Personal Influence, Collective Rationality, and Mass Political Action." *American Political Science Review* 83: 885–905.

Flam, Helena. 1994. *States and Anti-Nuclear Movements*. Edinburgh: Edinburgh University Press.

Freedman, David A. 2006. "Statistical Models for Causation: What Inferential Leverage Do They Provide?" *Evaluation Review* 30: 691–713.

    2010. "On Types of Scientific Inquiry: The Role of Qualitative Reasoning." Pp. 221–36 in *Rethinking Social Inquiry*, ed. Henry E. Brady and David Collier. Lanham, MD: Rowman and Littlefield.

Freudenburg, William R. 2004. "Can We Learn from Failure? Examining U.S. Experiences with Nuclear Repository Siting." *Journal of Risk Research* 7: 153–69.

Freudenburg, William R. and R. Gramling. 1993. "Socioenvironmental Factors and Development Policy: Understanding Opposition and Support for Offshore Oil." *Sociological Forum* 8: 341–64.

    1994. *Oil in Troubled Waters: Perceptions, Politics, and the Battle over Offshore Drilling*. Albany: State University of New York Press.

Gallagher, L., S. Ferreira, and F. Convery. 2008. "Host Community Attitudes toward Solid Waste Landfill Infrastructure Comprehension before Compensation." *Journal of Environmental Planning and Management* 51: 233–57.

Gamson, William A. 1968a. *Power and Discontent*. Homewood, IL: Dorsey.

    1968b. "Stable Unrepresentation in American Society." *American Behavioral Scientist* 12: 15–21.

    1990 [1975]. *The Strategy of Social Protest*. 2nd ed. Belmont, CA: Wadsworth.

Gamson, William A. and David S. Meyer. 1996. "Framing Political Opportunity." Pp. 275–90 in *Comparative Perspectives on Social Movements*, eds. Doug McAdam, John D. McCarthy, and Mayer Zald. New York: Cambridge University Press.

Ganz, Marshal. 2009. *Why David Sometimes Wins: Leadership, Organization and Strategy in the California Farm Worker Movement*. New York: Oxford University Press.

George, Alexander L. 1979. "Case Studies and Theory Development: The Method of Structured, Focused Comparison." Pp. 43–68 in *Diplomacy: New Approaches in History, Theory and Policy*, ed. Paul Gordon Lauren. New York: Free Press.

George, Alexander L. and Andrew Bennett. 2005. *Case Studies and Theory Development*. Cambridge, MA: MIT Press.

George, Alexander L. and Richard Smoke. 1974. *Deterrence in American Foreign Policy and Practice*. New York: Columbia University Press.

Giugni, Marco. 1998. "Was It Worth the Effort? The Outcomes and Consequences of Social Movements." *Annual Review of Sociology* 24: 371–93.

Giugni, Marco, Doug McAdam, and Charles Tilly, eds. 1999. *How Social Movements Matter*. Minneapolis: University of Minnesota Press.

Glassman, J. 2002. "From Seattle (and Ubon) to Bangkok: The Scales of Resistance to Corporate Globalization." *Environment and Planning D: Society and Space* 20: 513–33.

Goodwin, Jeff and Gabriel Hetland. 2010. "The Strange Disappearance of Capitalism from Social Movement Studies." Unpublished paper.

Goodwin, Jeff and James M. Jasper. 1999. "Caught in a Winding, Snarling Vine: The Structural Bias of Political Process Theory." *Sociological Forum* 14: 27–54.

Gould, Deborah B. 2009. *Moving Politics: Emotion and ACT UP's Fight against AIDS*. Chicago: University of Chicago Press.

Gould, Roger. 1991. "Multiple Networks and Mobilization in the Paris Commune, 1871." *American Sociological Review* 56: 716–29.

1993. "Collective Action and Network Structures." *American Sociological Review* 58: 182–96.

1995. *Insurgent Identities: Class, Community and Protest in Paris from 1848 to the Commune*. Chicago: University of Chicago Press.

Gramling, R. 1995. *Oil on the Edge: Offshore Development, Conflict, Gridlock*. Albany: State University of New York Press.

Gramling, R. and William R. Freudenburg. 1996. "Crude, Coppertone, and the Coast: Developmental Channelization and Constraint of Alternative Development Opportunities." *Society and Natural Resources* 9: 483–506.

2006. "Attitudes toward Offshore Oil Development: A Summary of Current Evidence." *Ocean and Coastal Management* 49: 20–46.

Grant, Don, Andrew Jones, and Albert Bergesen. 2002. "Organizational Size and Pollution: The Case of the U.S. Chemical Industry." *American Sociological Review* 67: 389–408.

Gusfield, Joseph R. 1970. *Protest, Reform and Revolt*. New York: John Wiley and Sons.

Gustafsson, Håkan and Stellan Vinthagen. 2011. "Law on the Move: Mapping the Dynamic Interactions of Law and Social Movements." Unpublished paper.

Gustaitis, Rasa. 2002. "The Jewel of Oxnard." *California Coast and Ocean* 18(2): 2–6, 32.

Hallett, Tim. 2010. "The Myth Incarnate: Recoupling Processes, Turmoil, and Inhabited Institutions in an Urban Elementary School." *American Sociological Review* 75: 52–74.

Hamilton, J. T. 1993. "Politics and Social Costs: Estimating the Impact of Collective Action on Hazardous Waste Facilities." *The RAND Journal of Economics* 24: 101–25.

Harvey, David. 2000. *Spaces of Hope*. Berkeley: University of California Press.

Hawes, K. 2004. "LNG Terminal Will Be an Asset." *Brazosport Facts*, May 18.

Herdt, Timm. 2005. "Governor Wants LNG Site off Oxnard." *Ventura County Star*, June 24.

Hobsbawm, Eric J. 1959. *Primitive Rebels: Studies in Archaic Forms of Social Movement in the 19th and 20th Centuries*. Manchester, UK: Manchester University Press.

1962. *The Age of Revolution: 1789–1848*. London: Weidenfeld and Nicholson.

Hoffer, Eric. 1951. *The True Believer: Thoughts on the Nature of Mass Movements*. New York: The New American Library.

Hunter, S. and K. M. Leyden. 1995. "Beyond NIMBY: Explaining Opposition to Hazardous Waste Facilities." *Policy Studies Journal* 23: 601–4.

Incantalupo, T. 2005. "Sound of Dissent over Gas Terminal." *Long Island Newsday*, January 18.

Ingram, Paul, Lori Qingyuan Yue, and Hayagreeva Rao. 2010. "Trouble in Store: Probes, Protests and Store Openings by Walmart: 1998–2007." *American Journal of Sociology* 116 (1): 53–92.

James, A. 2005. "LNG Fuels Heated Debate at Gathering of Lawmakers." *Press-Register*, August 1.

Jenkins, J. Craig. 1985. *The Politics of Insurgency: The Farm Worker Movement in the 1960s*. New York: Columbia University Press.

Jenkins, J. Craig and Charles Perrow. 1977. "Insurgency of the Powerless: Farm Workers' Movements (1946–1972)." *American Sociological Review* 42: 249–68.

Jenkins, J. Craig, David Jacobs, and Jon Agnone. 2003. "Political Opportunities and African-American Protest, 1948–1997." *American Journal of Sociology* 109: 277–303.

Johns, Paul. 2003. "Is There Life after Policy Streams, Advocacy Coalitions, and Punctuations: Using Exploratory Theory to Explain Policy Change?" *Policy Studies Journal* 31: 481–98.

Kay, Tamara. 2005. "Labor Transnationalism and Global Governance: The Impact of NAFTA on Transnational Labor Relationships in North America." *American Journal of Sociology* 111: 715–56.

Kimeldorf, Howard. 1999. *Battling for American Labor: Wobblies, Craft Workers, and the Making of the Union Movement*. Berkeley and Los Angeles: University of California Press.

King, Gary, Robert O. Keohane, and Sidney Verba. 1994. *Designing Social Inquiry: Scientific Inference in Qualitative Research*. Princeton, NJ: Princeton University Press.

Kingdon, John. 1995. *Agendas, Alternatives, and Public Policies*. 2nd ed. Boston: Little, Brown.

Kitschelt, Herbert. 1986. "Political Opportunity Structures and Political Protest: Anti-Nuclear Movements in Four Democracies." *British Journal of Political Science* 16: 57–85.

1995. *The Radical Right in Western Europe: A Comparative Analysis*. Ann Arbor: University of Michigan Press.

Klandermans, Bert and Dirk Oegema. 1987. "Potentials, Networks, Motivation, and Barriers: Steps toward Participation in Social Movements." *American Sociological Review* 52: 519–31.

Klapp, Orrin. 1969. *Collective Search for Identity*. New York: Holt, Rinehart, and Winston.

Koopmans, Ruud. 1993. "The Dynamics of Protest Waves: West Germany, 1965–1989." *American Sociological Review* 58: 637–58.

1995. *Democracy from Below: New Social Movements and the Political System in West Germany*. Boulder, CO: Westview Pres.

Kriesi, Hanspeter. 2004. "Political Context and Opportunity." Pp. 67–90 in *The Blackwell Companion to Social Movements*, ed. David A. Snow, Sarah A. Soule and Hanspeter Kriesi. Malden, MA: Blackwell Publishing.

Kriesi, Hanspeter, Ruud Koopmans, Jan Willem Duyvendak, and Marco G. Giugni. 1995. *New Social Movements in Western Europe*. Minneapolis: University of Minnesota Press.

Kunreuther, H. and J. Linnerooth. 1982. *Risk Analysis and Decision Processes: The Siting of Liquified Energy Facilities in Four Countries*. Berlin: Springer-Verlag.

Laetz, H. 2008. "Local LNG Foes Get Boost from Major Environmental Watchdog." *Malibu Surfside News*, January 3.

Lang, Kurt and Gladys Lang. 1961. *Collective Dynamics*. New York: Crowell.

Lasswell, Harold D. 1958. *Politics: Who Gets What, When, How*. New York: Meridian Books.

LeBon, Gustave. 1897. *The Crowd*. London: Unwin.

Lesbirel, S. H. 1998. *NIMBY Politics in Japan: Energy Siting and the Management of Environmental Conflict*. Ithaca, NY: Cornell University Press.

Lesbirel, S. Hayden and Daigee Shaw, eds. 2005. *Managing Conflict in Facility Siting: An International Comparison*. Northhampton, MA: Edward Elgar.

Lichbach, Mark Irving. 2008. "Modeling Mechanisms of Contention: MTT's Positivist Contribution." *Qualitative Sociology* 31: 345–54.

Lipsky, Michael. 1968. "Protest as a Political Resource." *American Political Science Review* 62: 1144–58.

1970. *Protest in City Politics*. Chicago: Rand McNally.

LNG Long Haul. 2009. *Bangor Daily News*. Bangor, ME, October 2.

Lober, D. J. 1995. "Why Protest? Public Behavioral and Attitudinal Response to Siting a Waste Disposal Facility." *Policy Studies Journal* 23: 499–519.

Lounsbury, Michael. 2005. "Institutional Variation in the Evolution of Social Movements: Competing Logics and the Spread of Recycling Advocacy Groups." Pp. 73–95 in *Social Movements and Organization Theory*, ed. Gerald F. Davis, Doug McAdam, W. Richard Scott, and Mayer N. Zald. New York: Cambridge University Press.

Luders, Joseph E. 2010. *The Civil Rights Movement and the Logic of Social Change*. New York: Cambridge University Press.

Lystra, T. 2009. "State Agency, Riverkeeper Sue to Halt LNG Project." *The Daily News Online*, February 13.

Mansbridge, Jane. 1986. *Why We Lost the ERA*. Chicago: University of Chicago Press.

McAdam, Doug. 1983. "Tactical Innovation and the Pace of Insurgency." *American Sociological Review* 48: 735–54.

1986. "Recruitment to High-Risk Activism: The Case of Freedom Summer." *American Journal of Sociology* 92: 64–90.

1988. *Freedom Summer*. New York: Oxford University Press.

1995. "'Initiator' and 'Spin-Off' Movements: Diffusion Processes in Cycles of Protest." Pp. 217–39 in *Repertoires and Cycles of Collective Action*, ed. Mark Traugott. Durham, NC: Duke University Press.

1999 [1982]. *Political Process and the Development of Black Insurgency, 1930–1970*. Chicago: University of Chicago Press.

2008. "Methods for Modeling Mechanisms of Contention." *Qualitative Sociology* 31: 307–32.

McAdam, Doug and Ronnelle Paulsen. 1993. "Specifying the Relationship between Social Ties and Activism." *American Journal of Sociology* 99: 640–67.

McAdam, Doug and Dieter Rucht. 1994. "The Cross National Diffusion of Movement Ideas." *Annals of the American Academy of Political and Social Science* 528: 36–59.

McAdam, Doug and David A. Snow. 1997. *Social Movements: Readings on Their Emergence, Mobilization and Dynamics*. Los Angeles: Roxbury Publishing Company.

McAdam, Doug and Yang Su. 2002. "The War at Home: Antiwar Protests and Congressional Voting, 1965–1972." *American Sociological Review* 67: 696–721.

McAdam, Doug, John D. McCarthy, and Mayer N. Zald. 1988. "Social Movements." Pp. 695–737 in *Handbook of Sociology*, ed. Neil J. Smelser. Beverly Hills, CA: Sage Publications.

1996. *Comparative Perspectives on Social Movements*. Cambridge and New York: Cambridge University Press.

McAdam, Doug, Sidney Tarrow, and Charles Tilly. 2001. *Dynamics of Contention*. New York: Cambridge University Press.

McAdam, Doug, Robert J. Sampson, Simon Weffer, and Heather MacIndoe. 2005. "'There Will Be Fighting in the Streets': The Distorting Lens of Social Movement Theory." *Mobilization* 10: 1–18.

McAdam, Doug, Hilary S. Boudet, Jenna Davis, Ryan J. Orr, W. Richard Scott, and Ray E. Levitt. 2010. "'Site Fights': Explaining Opposition to Pipeline Projects in the Developing World." *Sociological Forum* 25: 401–27.

McCammon, Holly J., Karen E. Campbell, Ellen M. Granberg, and Christine Mowery. 2001. "How Movements Win: Gendered Opportunity Structures and U.S. Women's Suffrage Movements, 1866–1919." *American Sociological Review* 66: 49–70.

McCammon, Holly J., Courtney Sanders Muse, Harmony D. Newman, and Teresa M. Terrell. 2007. "Movement Framing and Discursive Opportunity Structures: The Political Successes of the U.S. Women's Jury Movements." *American Sociological Review* 72: 725–49.

McCann, Michael. 1994. *Rights at Work: Pay Equity Reform and the Politics of Legal Mobilization*. Chicago: University of Chicago Press.

McCarthy, John D. and Mayer N. Zald. 1973. *The Trend of Social Movements in America: Professionalization and Resource Mobilization*. Morristown, NJ: General Learning Press.

1977. "Resource Mobilization and Social Movements: A Partial Theory." *American Journal of Sociology* 82: 1212–41.

McCarthy, John D., Clark McPhail, and Jackie Smith. 1996. "Images of Protest: Estimating Selection Bias in Media Coverage of Washington Demonstrations, 1982 and 1991." *American Sociological Review* 61: 478–99.

McVeigh, Rory. 2009. *The Rise of the Ku Klux Klan: Right-Wing Movements and National Politics*. Minneapolis: University of Minnesota Press.

Meyer, David S. 1990. *A Winter of Discontent: The Nuclear Freeze and American Politics*. New York: Praeger.

Meyer, David S. and Debra C. Minkoff. 2004. "Conceptualizing Political Opportunity." *Social Forces* 82 (4): 1457–92.

Meyer, David S. and Sidney Tarrow, eds. 1998. *The Social Movement Society*. Lanham, MD: Rowman and Littlefield.

Meyer, John W., John Boli, George Thomas, and Francisco O. Ramirez. 1997. "World Society and the Nation State." *American Journal of Sociology* 103: 1444–81.

Meyer, John W., David John Frank, Ann Hironaka, Evan Schofer, and Nancy Tuma. 1997. "The Structuring of a World Environmental Regime, 1870–1990." *International Organization* 51: 623–51.

Miller, Byron. 2000. *Geography and Social Movements*. Minneapolis: University of Minnesota Press.

2004. "Spaces of Mobilization: Transnational Social Movements." Pp. 223–46 in *Spaces of Democracy: Geographical Perspectives on Citizenship, Participation and Representation*, ed. C. Barnett and M. Low. London: Sage Publications.

Mills, C. Wright. 1959. *The Power Elite*. New York: Oxford University Press.

Minkoff, Debra. 1993. "The Organization of Survival: Women's and Racial-Ethnic Voluntarist and Activist Organizations, 1955–1985." *Social Forces* 71 (4): 887–908.

1995. *Organizing for Equality: The Evolution of Women's and Racial-Ethnic Organizations in America, 1955–1985*. New Brunswick, NJ: Rutgers University Press.

1997. "The Sequencing of Social Movements." *American Sociological Review* 62: 779–99.

Mische, Ann. 2008. *Partisan Publics*. Princeton, NJ: Princeton University Press.

Molotch, Harvey, William R. Freudenburg, and K. E. Paulsen. 2000. "History Repeats Itself, But How? City Character, Urban Tradition, and the Accomplishment of Place." *American Sociological Review* 65: 791–823.

Morris, Aldon D. 1981. "Black Southern Sit-in Movement: An Analysis of Internal Organization." *American Sociological Review* 46: 744–67.

1984. *The Origins of the Civil Rights Movement*. New York: Free Press.

Musick, Mark A. and John Wilson. 2008. *Volunteers: A Social Profile*. Bloomington: Indiana University Press.

Muthukumara, Mani and David Wheeler. 1998. "In Search of Pollution Havens: Dirty Industry in the World Economy, 1960–1996." *Journal of Environment and Development* 7: 215–47.

Myers, Daniel J. 2000. "The Diffusion of Collective Violence: Infectiousness, Susceptibility, and Mass Media Networks." *American Journal of Sociology* 106: 173–208.

Nagel, Joanne. 1996. *American Indian Ethnic Renewal: Red Power and the Resurgence of Identity*. New York: Oxford University Press.

National Marine Fisheries Service. 2005. National Marine Fisheries Service Comment Letter on Gulf Landing Final EIS. Silver Spring, MD.

Nie, Norman H. and Sidney Verba. 1975. "Political Participation." Pp. 1–74 in *Handbook of Political Science*, vol. 4, *Nongovernmental Politics*, ed. F. I. Greenstein and Nelson Polsby. Reading, PA: Addison-Wesley.

No Way On OceanWay. 2010. "10 Reasons Why the OceanWay LNG Project Is Wrong for Southern California." LNG Watch, April 1, 2008, http://lngwatch. live.radicaldesigns.org/downloads/Woodside%20PDF_2pager_04-01-08.pdf (accessed August 1, 2010).

Nuclear Regulatory Commission. 2005. Draft EIS for an Early Site Permit (ESP) at the Grand Gulf ESP site. June 28, 67–9.

Oberschall, Anthony. 1989. "The 1960 Sit-ins: Protest Diffusion and Movement Takeoff." *Research in Social Movements, Conflict and Change* 11: 31–53.

O'Brien, Kevin J. and Lianjiang Li. 2006. *Rightful Resistance in Rural China*. New York: Cambridge University Press.

Oliver, Pamela E. 1984. "'If You Don't Do It, Nobody Else Will': Active and Token Contributors to Local Collective Action." *American Sociological Review* 49: 601–10.

Olson, Mancur, Jr. 1965. *The Logic of Collective Action*. Cambridge, MA: Harvard University Press.

Olzak, Susan, Maya Beasley, and Johan Olivier. 2003. "The Impact of State Reforms on Protest against Apartheid in South Africa." *Mobilization* 8: 27–50.

Paige, Jeffrey M. 1975. *Agrarian Revolution*. New York: Free Press.
    1997. *Coffee and Power: Revolution and the Rise of Democracy in Central America*. Cambridge, MA: Harvard University Press.

Parsa, Misagh. 1989. *Social Origins of the Iranian Revolution*. New Brunswick, NJ: Rutgers University Press.

Parsons, Talcott. 1951. *The Social System*. Glencoe, IL: Free Press.
    1971. *The System of Modern Societies*. Englewood Cliffs, NJ: Prentice-Hall.

Passy, Florence. 2001. "Socializing, Connecting, and the Structural Agency Gap: A Specification of the Impact of Networks on Participation in Social Movements." *Mobilization* 6: 173–92.
    2003. "Social Networks Matter, But How?" Pp. 21–48 in *Social Movements and Networks*, ed. Mario Diani and Doug McAdam. Oxford: Oxford University Press.

Pedriana, Nicholas. 2006. "From Protective to Equal Treatment: Legal Framing Processes and Transformation of the Women's Movement in the 1960s." *American Journal of Sociology* 111: 1718–61.

Piller, C. 1991. *The Fail-Safe Society: Community Defiance and the End of American Technological Optimism*. New York: Basic Books.

Pitcher, Brian, Robert Hamblin, and Jerry Miller. 1977. "The Diffusion of Collective Violence." *American Sociological Review* 43: 23–35.

Piven, Frances Fox and Richard Cloward. 1977. *Poor People's Movements: Why They Succeed, How They Fail*. New York: Vintage Books.

Powell, J. 2004. "Done Deal: LNG Plant Is Coming, ExxonMobil Facility Will Be in Ingleside." *Corpus Christi Caller-Times*, January 16.

Ragin, Charles C. 1987. *The Comparative Method*. Berkeley: University of California Press.

2000. *Fuzzy-Set Social Science*. Chicago: University of Chicago Press.

2004. "Turning the Tables: How Case-Oriented Research Challenges Variable-Oriented Research." Pp. 125–41 in *Rethinking Social Inquiry: Diverse Tools, Shared Standards*, ed. Henry E. Brady and David Collier. Lanham, MD: Rowman and Littlefield.

2008. *Redesigning Social Inquiry: Fuzzy Sets and Beyond*. Chicago: University of Chicago Press.

Rao, Hayagreeeva, Philippe Monin, and Rodolphe Durand. 2003. "Institutional Change in Toque Ville: Nouvelle Cuisine as an Identity Movement in French Gastronomy." *American Journal of Sociology* 108: 795–843.

Reynolds, M. 2004. "Carcieri Opposes LNG Plan." *Providence Journal*, September 30.

Rojas, Fabio. 2006. "Social Movement Tactics, Organizational Change and the Spread of African-American Studies." *Social Forces* 84: 2139–58.

2007. *From Black Power to Black Studies: How a Radical Social Movement Became an Academic Discipline*. Baltimore, MD: Johns Hopkins University Press.

Rootes, Christopher. 2003. *Environmental Protest in Western Europe*. Oxford: Oxford University Press.

2009a. "Environmental Movements and Campaigns against Waste Infrastructure in the United States." *Environmental Politics* 18: 835–50.

2009b. "Environmental Movements, Waste and Waste Infrastructure: An Introduction." *Environmental Politics* 18: 817–34.

Rucht, Dieter. 1999. "Linking Organization and Mobilization: Michels' 'Iron Law of Oligarchy' Reconsidered." *Mobilization* 4: 151–70.

Rudé, George. 1959. *The Crowd in the French Revolution*. Oxford: Oxford University Press.

1964. *The Crowd in History*. New York: Wiley.

Rule, James and Charles Tilly. 1972. "1830 and the Unnatural History of Revolution." *Journal of Social Issues* 28: 49–76.

Rupp, Leila J. and Verta Taylor. 1987. *Survival in the Doldrums*. New York: Oxford University Press.

Sabatier, Paul A., ed. 2007. *Theories of the Policy Process*. 2nd ed. Boulder: CO: Westview Press.

Sabatier, Paul A. and H. Jenkins-Smith. 1988. "An Advocacy Coalition Model of Policy Change and the Role of Policy Orientated Learning Therein." *Policy Sciences* 21: 129–68.

1993. *Policy Change and Learning: An Advocacy Coalition Approach*. Boulder, CO: Westview Press.

Sabatier, Paul A. and Neil Pelkey. 1987. "Incorporating Multiple Actors and Guidance Instruments into Models of Regulatory Policymaking: An Advocacy Coalition Framework." *Administration and Society* 19: 236–63.

Schively, C. 2007. "Understanding the NIMBY and LULU Phenomena: Reassessing Our Knowledge Base and Informing Future Research." *Journal of Planning Literature* 21: 255–66.

Schlager, E. 1995. "Policy Making and Collective Action: Defining Coalitions within the Advocacy Coalition Framework." *Policy Sciences* 28: 242–70.

Schleifstein, M. 2005. "LNG Plant's Impact Revised: Redfish Toll Estimate Drastically Lowered." *The Times-Picayune*, February 7, 2005.

Schwartz, Michael. 1976. *Radical Protest and Social Structure: The Southern Farmers' Alliance and Cotton Tenancy, 1880–1890*. New York: Academic Press.

Sewell, William H., Jr. 1992. "A Theory of Structure: Duality, Agency and Transformation." *American Journal of Sociology* 98: 1–29.

1996. "Historical Events As Transformations of Structure: Inventing Revolution at the Bastille." *Theory and Society* 25: 841–81.

Shaw, S. 2004. Letter-to-the-Editor: Natural Gas Issue Critical." *Brazosport Facts*, August 30.

Shepard, B. and R. Hayduk, eds. 2002. *From ACT UP to the WTO: Urban Protest and Community Building in the Era of Globalization*. New York: Verso.

Sherman, Daniel J. 2011a. "Critical Mechanisms for Critical Masses: Exploring Opposition to Variation in Low-Level Radioactive Waste Sight Proposals." *Mobilization* 16: 721–43.

2011b. "Not Here, Not There, Not Anywhere: Politics, Social Movements, and the Disposal of Low-Level Radioactive Waste." Washington, DC and London: RFF Press.

Shorter, Edward and Charles Tilly. 1974. *Strikes in France, 1830–1968*. New York: Cambridge University Press.

Sierra Club. 2009. "Energy Resources Policy," May 16, 2009, http://www.sierra-club.org/policy/conservation/energy.pdf (accessed August 1, 2010).

Sierra Club California. 2005. "Sierra Club California LNG Decision," March 22, 2005, http://angeles.sierraclub.org/hvtf/lng/LNG_files/SCRCC_decision.html (accessed August 1, 2010).

Sierra Club Oregon. 2010. "Stop Liquefied Natural Gas," July 26, 2010, http://oregon.sierraclub.org/goals/lng.asp (accessed August 1, 2010).

Simon, David R. 2000. "Corporate Environmental Crimes and Social Inequality: New Directions for Environmental Research." *American Behavioral Scientist* 43: 633–45.

Skocpol, Theda. 1979. *States and Social Revolutions*. New York: Cambridge University Press.

1985. "Bringing the State Back In: Strategies of Analysis in Current Research." Pp. 3–37 in *Bringing the State Back In*, ed. Peter Evans, Dietrich Rueschemeyer, and Theda Skocpol. Cambridge: Cambridge University Press.

Slaughter, Sheila. 1997. "Class, Race, Gender and the Construction of Post-Secondary Curricula in the United States: Social Movement,

Professionalization and Political Economic Theories of Curricular Change."
Journal of Curriculum Studies 29: 1–30.

Slovic, P. 1987. "Perception of Risk." *Science* 236: 280–5.

Smelser, Neil. 1962. *Theory of Collective Behavior*. New York: Free Press.

Smith, Christian. 1991. *The Emergence of Liberation Theology*. Chicago: University of Chicago Press.

Smith, M. 2005. "Incentives Key to Future of Chemical Industry." *Brazosport Facts*, August 14.

Snow, David A. and Robert D. Benford. 1988. "Ideology, Frame Resonance, and Participant Mobilization." Pp. 197–217 in *From Structure to Action: Social Movement Participation across Cultures*, ed. Bert Klandermans, Hanspeter Kriesi, and Sidney Tarrow. Greenwich, CT: JAI Press.

  1992. "Master Frames and Cycles of Protest." Pp. 133–55 in *Frontiers in Social Movement Theory*, ed. Aldon D. Morris and Carol McClurg Mueller. New Haven, CT: Yale University Press.

Snow, David A., Sarah A. Soule, and Hanspeter Kriesi, eds. 2004. *The Blackwell Companion to Social Movements*. Malden, MA: Blackwell.

Snow, David A., Louis A. Zurcher, and Sheldon Ekland-Olson. 1980. "Social Networks and Social Movements." *American Sociological Review* 45: 787–801.

Snow, David A., E. Burke Rochford, Steven K. Worden, and Robert D. Benford. 1986. "Frame Alignment Processes, Micromobilization, and Movement Participation." *American Sociological Review* 51: 464–81.

Snyder, David and Charles Tilly. 1972. "Hardship and Collective Violence in France, 1830–1960." *American Sociological Review* 37: 520–32.

Soule, Sarah A. 1995. "The Student Anti-Apartheid Movement in the United States: Diffusion of Protest Tactics ad Policy Reform." PhD diss., Department of Sociology, Cornell University, Ithaca, NY.

  1997. "The Student Divestment Movement in the United States and Tactical Diffusion: The Shantytown Protest." *Social Forces* 75: 855–83.

Soule, Sarah A. and Jennifer Earl. 2001. "The Enactment of State-Level Hate Crime Law in the United States: Intrastate and Interstate Factors." *Sociological Perspectives* 44: 281–305.

Soule, Sarah A. and Susan Olzak. 2004. "When Do Movements Matter? The Politics of Contingency and the Equal Rights Amendment." *American Sociological Review* 69: 473–97.

Soule, Sarah A., Doug McAdam, John McCarthy, and Yang Su. 1999. "Protest Events: Causes or Consequences of State Action? The U.S. Women's Movement and Federal Congressional Activity." *Mobilization* 4: 239–56.

Staggenborg, Suzanne. 2008. "Seeing Mechanisms in Action." *Qualitative Sociology* 31: 341–44.

Strang, David and Sarah A. Soule. 1998. "Diffusion in Organizations and Social Movements: From Hybrid Corn to Poison Pills." *Annual Review of Sociology* 24: 265–90.

Steedly, Homer R. and John W. Foley. 1979. "The Success of Protest Groups: Multivariate Analysis." *Social Science Research* 8: 1–15.

Steinberg, Marc W. 1994. "The Dialogue of Struggle: The Contest over Ideological Boundaries in the Case of London Silk Weavers in the Early Nineteenth Century." *Social Science History* 18: 505–41.

1999. "The Talk and Balk Talk of Collective Action: A Dialogic Analysis of Repertoires of Discourse among Nineteenth-Century English Cotton Spinners." *American Journal of Sociology* 105: 736–80.

Stevens, Mitchell L. 2001. *Kingdom of Children: Culture and Controversy in the Homeschooling Movement.* Princeton, NJ: Princeton University Press.

Swidler, Ann. 1986. "Culture in Action: Symbols and Strategies." *American Sociological Review* 51: 273–86.

Tarde, Gabriel. 1903. *The Laws of Imitation.* New York: Holt.

Tarrow, Sidney. 1967. *Peasant Communism in Southern Italy.* New Haven, CT: Yale University Press.

1983. "Struggling to Reform: Social Movements and Policy Change during Cycles of Contention." Western Societies Paper 15. Ithaca, NY: Cornell University.

1989. *Democracy and Disorder: Protest and Politics in Italy, 1965–1975.* Oxford: Clarendon Press.

1998. *Power in Movement.* Cambridge: Cambridge University Press.

2005. *The New Transnational Activism.* New York: Cambridge University Press.

2010. "Bridging the Quantitative-Qualitative Divide." Pp. 101–10 in *Rethinking Social Inquiry,* ed. Henry E. Brady and David Collier. Lanham, MD: Rowman and Littlefield.

Tarrow, Sidney and Doug McAdam. 2005. "Scale Shift in Transnational Contention." Pp. 121–47 in *Transnational Protest and Global Activism,* ed. Donatella della Porta and Sidney Tarrow. Boulder, CO: Rowman and Littlefield.

Taylor, Verta. 1996. *Rock-a-by Baby: Feminism, Self-Help and Postpartum Depression.* New York: Routledge.

Thompson, E. P. 1963. *The Makings of the English Working Class.* New York: Vintage.

Tilly, Charles. 1964. *The Vendee.* Cambridge, MA: Harvard University Press.

1969. "Collective Violence in European Perspective." Pp. 4–45 in *Violence in America: Historical and Comparative Perspectives,* ed. Hugh Davis Graham and Ted Robert Gurr. Washington, DC: U.S. Government Printing Office.

1978. *From Mobilization to Revolution.* Reading, MA: Addison-Wesley.

Tilly, Charles and Lesley J. Wood. 2003. "Contentious Connections in Great Britain, 1828–34." Pp. 147–72 in *Social Movements and Networks,* ed. Mario Diani and Doug McAdam. Oxford: Oxford University Press.

Tilly, Charles, Louise Tilly, and Richard Tilly. 1975. *The Rebellious Century, 1830–1930.* Cambridge, MA: Harvard University Press.

Tindall, David B. 2004. "Social Movement Participation over Time: An Ego-Network Approach to Micro-Mobilization." *Sociological Focus* 37: 163–84.

True, James L., Bryan D. Jones, and Frank R. Baumgartner. 2007. "Punctuated-Equilibrium Theory: Explaining Stability and Change in Public Policymaking." Pp. 155–187 in *Theories of the Policy Process.* 2nd ed., ed. Paul A. Sabatier. Boulder, CO: Westview Press.

Tsutsui, Kiyoteru. 2004. "Global Civil Society and Ethnic Social Movements in the Contemporary World." *Sociological Forum* 19: 63–87.

Turner, Ralph and Lewis Killian. 1957. *Collective Behavior*. Englewood Cliffs, NJ: Prentice-Hall.

Tversky, Amos and Donald Kahneman. 1974. "Judgment under Uncertainty." *Science* 185: 1124–31.

U.S. Department of Interior. 2008. "Case Study: Texas Operations Contractor Alliance for Safety at Dow Facility in Freeport, Texas." Washington, DC: U.S. Occupational Safety and Health Administration.

Vallejo for Community Planned Renewal. 2010. "Top 10 Reasons Why LNG Is Wrong for Vallejo." maggdog communications, January 8 2003, http://www.vallejocpr.org/background/top10.html (accessed August 1, 2010).

Vajjhala, S. P. and P. S. Fischbeck. 2006. *Quantifying Siting Difficulty: A Case Study of U.S. Transmission Line Siting*. Washington, DC: Resources for the Future.

Vasi, Ion Bogdan. 2011. *Winds of Change: The Environmental Movement and the Global Development of the Wind Energy Industry*. Oxford and New York: Oxford University Press.

Verba, Sidney, Kay Lehman Schlozman, and Henry E. Brady. 1995. *Voice and Equality*. Cambridge, MA: Harvard University Press.

Vogus, Timothy J. and Gerald F. Davis. 2005. "Elite Mobilization for Antitakeover Legislation, 1982–1990. Pp. 96–121 in *Social Movements and Organization Theory*, ed. Gerald F. Davis, Doug McAdam, W. Richard Scott and Mayer N. Zald. New York: Cambridge University Press.

Walder, Andrew. 2009a. "Political Sociology and Social Movements." *Annual Review of Sociology* 35: 393–412.

2009b. *Fractured Rebellion*. Cambridge, MA: Harvard University Press.

Walsh, Edward J., Rex H. Warland, and D. C. Smith. 1997. *Don't Burn It Here: Grassroots Challenges to Trash Incinerators*. University Park: Pennsylvania State University Press.

Weiss, R. S. 1994. *Learning from Strangers: The Art and Method of Qualitative Interview Studies*. New York: The Free Press.

Whitmore, W. D., V. K. Baxter, and S. L. Laska. 2008. "A Critique of Offshore Liquefied Natural Gas (LNG) Terminal Policy." *Ocean and Coastal Management* 52: 10–16.

Wise, J. 2003. "Police Jury Given Report on New Cheniere LNG Plant." *Parish Pilot*, May 8.

Wood, Elisabeth Jean. 2003. *Insurgent Collective Action and Civil War in El Salvador*. Cambridge: Cambridge University Press.

Yashar, Deborah J. 2005. *Contesting Citizenship in Latin America: The Rise of Indigenous Movements and the Postliberal Challenge*. New York: Cambridge University Press.

Zhao, Dingxin. 1998. "Ecologies of Social Movements: Student Mobilization during the 1989 Prodemocracy Movement in Beijing." *American Journal of Sociology* 103: 1493–1529.

Index